BIRKHÄUSER

Progress in Mathematical Physics
Volume 57

Mikhail V. Volkenstein

Entropy and Information

Translated by Abe Shenitzer and Robert G. Burns

Birkhäuser
Basel · Boston · Berlin

Author: Mikhail V. Volkenstein (1912 – 1992)

Translators:

Abe Shenitzer
4 Oakdale Street
Jamaica Plain
MA 02130
USA

Robert G. Burns
Dept. of Math. and Stat.
York University
4700 Keele Street
Toronto M3J 1P3
Canada

2000 Mathematics Subject Classification: 94A17, 68P30, 33B10, 11B73

Library of Congress Control Number: 2009931264

Bibliographic information published by Die Deutsche Bibliothek.
Die Deutsche Bibliothek lists this publication in the Deutsche Nationalbibliografie; detailed bibliographic data is available in the Internet at http://dnb.ddb.de

ISBN 978-3-0346-0077-4 Birkhäuser Verlag AG, Basel · Boston · Berlin

Basel · Boston · Berlin
P.O. Box 133, CH-4010 Basel, Switzerland
Part of Springer Science+Business Media
Printed on acid-free paper produced from chlorine-free pulp. TCF ∞
Printed in Germany

This book is a translation of the book "Энтропия и информация" written by Mikhail V. Volkenstein (М.В. Волькенштейн) and was published in the book series "Problems of Science and Technical Progress" by Москва Наука (Nauka Publishers) in 1986.

ISBN 978-3-0346-0077-4
9 8 7 6 5 4 3 2 1

e-ISBN 978-3-0346-0078-1
www.birkhauser.ch

Contents

About the Author

Entropy formed a central theme of the scientific work of Mikhail Vladimirovich Volkenstein, especially during the last period of his life. Initially, from about 1933 to 1948, his scientific interests lay more in spectroscopy; during this period, his most significant work concerned the intensities of molecular oscillation spectra. In the next period of his life, from around 1948 to 1960, he turned his attention to the new field of macromolecules, becoming one of the most important contributors to the statistical theory of such molecules. In the 1960s and 70s his interests turned naturally to questions from molecular biology and molecular biophysics. It is to him that we owe much of our understanding of biomolecular conformation; in particular, his work shed light on the relations between the chemical structure of polymers and their plastic and elastic properties, and on the reactivity of biomolecules. Towards the end of the 1970s, Mikhail Vladimirovich set out to explore the deep questions of the origin of life, and so began his investigations of the complex of problems linking "entropy, information, and life". It was at this time that I first met him, in Poland; his personality and his erudition made a profound impression on me. He subsequently attended several of our conferences in Berlin on the theme "Irreversible processes and self-organization", where he met Ilya Prigogine and became friends with him. I remember clearly the wonderful conversations about science and culture, full of controversy, between these two brilliant polyhistors, ranging from the dancing god Shiva to African sculpture, carried out now in one language,

Figure 1: Mikhail Vladimirovich Volkenstein during a visit to Berlin in 1981. (Please note that he was not a heavy smoker.)

now in some other. At that time Mikhail Vladimirovich was profoundly impressed by Manfred Eigen's work on the evolution of biological macromolecules and the origins of biological information, and began corresponding with him. In Moscow, however, Mikhail Vladimirovich was then considered a dissident and not permitted to visit Western countries. I recall how proud he was of a birthday present from Manfred Eigen, a tape of a piano recital performed by Eigen himself. Only towards the end of the 1980s was Mikhail Vladimirovich able to accept invitations from his numerous friends to travel with his wife to the West. I remember in particular his mentioning, apropos of a visit to Spain, how his wife had fought there a half-century earlier on the barricades.

I now turn to the present book, written by Mikhail Vladimirovich in the 1980s. The book is a brilliant essay covering a great many topics, ranging from the early history of thermodynamics to the complexity and value of information. I am certain that the reader will receive great enjoyment from this masterpiece of popular science.

Werner Ebeling
(Editor of the German translation
of *Entropy and Information*)

Translators' note

Our special thanks go to the following: Edwin F. Beschler for arranging publication and checking the typescript, Werner Ebeling for providing a biographical foreword about the author and a photograph of him, and also for correcting the spelling of some names, and Maria Volkenstein, the author's daughter, who, together with Aleksandr Y. Mozdakow and Vladimir Sychev of the publishing house Nauka, helped in obtaining the publication rights. We also thank Marc Herbstritt of Birkhäuser for handling the publication process so efficiently and amicably, and Nina Iukhoveli for help with Russian. The poetical epigraphs at the beginning of each chapter were originally rhyming; for the sake of accuracy, no attempt was made to produce rhyming English translations (with two exceptions), although we did try for some degree of rhythm. The second translator (Burns) acknowledges with gratitude the assistance of staff members of the departments of mathematics (and statistics) at York University, Toronto, and The University of Queensland, Brisbane.

The footnotes have, with rare exception, been added by the second translator, who takes full responsibility for them.

Preface

This is just...entropy,
he said, thinking that
this explained everything,
and he repeated the
strange word a few times.
Karel Čapek[1], "Krakatit"

This "strange word" denotes one of the most basic quantities of the physics of heat phenomena, that is, of thermodynamics. Although the concept of entropy did indeed originate in thermodynamics, it later became clear that it was a more universal concept, of fundamental significance for chemistry and biology, as well as physics.

Although the concept of energy is usually considered more important and easier to grasp, it turns out, as we shall see, that the idea of entropy is just as substantial—and moreover not all that complicated. We can compute or measure the quantity of energy contained in this sheet of paper, and the same is true of its entropy. Furthermore, entropy has remarkable properties. Our galaxy, the solar system, and the biosphere all take their being from entropy, as a result of its transference to the surrounding medium. There is a surprising connection between entropy and information, that is, the total intelligence communicated by a message. All of this is expounded in the present book, thereby conveying information to the reader and decreasing his entropy; but it is up to the reader to decide how valuable this information might be.

The second half of the 20th century is notable for the creation and development of complex areas of science of the greatest importance not only for the natural sciences and technology, but also for the humanities. Such are cybernetics, information theory, and synergetics. Although these terms did not exist fifty years ago[2], they now turn up constantly. In all three of these disciplines the concepts of entropy and information are absolutely indispensable, so that without them it is not possible to grasp the true essence of modern science. The final chapters of the

[1]Karel Čapek (1890–1938), Czech playwright and novelist. Inventor of the word "robot" in its present sense, in his play *RUR*.

[2]Note that the original work appeared in 1986. *Trans.*

book contain brief, and of necessity incomplete, expositions of synergetics and information theory. The aim of the present account is to bring these new disciplines to the reader's attention, and introduce him or her to the circle of related ideas.

I wish to thank M.I. Kaganov, Yu.L. Klimontovich, Ya.A. Smorodinskiĭ, and W. Ebeling for their useful comments on the manuscript.

M.V. Volkenstein

Chapter 1

"Reflections on the motive power of fire ..."

Black coal burns in the fire box,
Water turns into steam,
And the piston moves stubbornly
Back and forth, back and forth.

The hot steam condenses,
Turning the heavy wheels.
The engine-driving fire
Presents several riddles.

Sadi Carnot

A little book of only 45 pages, titled "Reflections on the motive power of fire, and on machines capable of developing that power"[1], appeared in Paris in 1824. Its author was S. Carnot, according to the title page a former student at the Polytechnic.

At that time Sadi Carnot was 28. He was the son of Lazare Carnot, who had been given the honorific title of "organizer of victory" for his services in the wars of revolutionary France. L. Carnot was a prominent military and political activist, an engineer, and an excellent mathematician; however the son's little memoir (as scientific papers were then called) was to outweigh in significance the works of the father.

This was S. Carnot's only publication. He died eight years later, in 1832, at the age of 36, a critical age for genius. Raphael, Mozart, and Pushkin died at about the same age.

[1]Sadi Carnot. *Réflexions sur la puissance motrice du feu.* 1824.

What was the state of physics at the time? Theoretical mechanics, whose foundations were discovered by the great Newton, seemed—with good reason—to have been perfected by Laplace, Lagrange, and other scientists. In optics there was a continuing struggle between the corpuscular and wave theories of light, though the final victory of the latter was well under way. In electricity and magnetism important discoveries had been made (by Ampère, Oersted, and Ohm), and the decisive discoveries of Faraday were imminent. Many properties of gases had been investigated in detail, and, thanks to the work of Dalton, atomic theory held sway in chemistry. In short, full-scale science—in particular physics—was advancing rapidly.

However, there was one important area, the physics of heat phenomena, that was still awaiting its Newton. There were two reasons for this delay. Firstly, steam engines—heat machines—had already infiltrated technology, and secondly, the essential features of heat phenomena turned out to be fundamentally different from those of mechanics.

As is clear from its title, Carnot's memoir relates directly to heat engines. The action of these machines prompted Carnot to think about heat and its transformation into work.

This is a relatively rare case in the history of science. Usually science develops independently of technology, following its own internal logic. Faraday, and after him Maxwell, were not thinking of the dynamo when they created the theory of the electromagnetic field. Maxwell, and then Hertz, were not led to their discovery of electromagnetic waves—first theoretically and then experimentally—by examining the functioning of radios. Sooner or later significant physics finds important practical applications.

Caloric

It was a different story for heat. The steam engine was invented independently of any theory. The story has it that James Watt's noticing how the lid of a boiling kettle bobbed up and down led him to the invention of the steam engine. Be that as it may, prior to Carnot's paper there existed no theory of heat whatsoever, and the important question of the mechanical efficiency of steam engines remained open. At that time it was difficult to construct a theory of heat; its nature was largely mysterious. There were two hypotheses. The first of these was based on the notion of the "caloric", a weightless, invisible fluid that when added to a body caused its temperature to rise and was capable of changing its state. Thus was the English physicist Joseph Black led to write down the following equations:

$$\text{ice} + \text{caloric} = \text{water},$$
$$\text{water} + \text{caloric} = \text{steam}.$$

It was thought that caloric was contained in a gas much like juice in an orange. Squeeze an orange and you get orange juice. Compress a gas and caloric oozes out, that is, it heats up.

The caloric theory was studied—without much success—by the famous revolutionary Jean-Paul Marat, who published an article on it in 1780.

The caloric hypothesis stood in opposition to the kinetic theory. In the mid-18th century M.V. Lomonosov and D. Bernoulli, in the course of propounding various arguments against the theory of the caloric, suggested a kinetic hypothesis. Lomonosov conjectured that "the cause of heat consists in the internal rotational motion" of the particles of the body, and that "the particles of hot bodies must spin faster and those of cooler bodies slower".

The observations of Benjamin Thompson (Count Rumford) decisively refuted the caloric hypothesis. In 1798 Rumford measured the amount of heat produced by drilling metal. Since the heat capacity of the shavings remained the same as that of the original metal, whence came the additional caloric? Rumford concluded that the source of heat must indeed be kinetic.[2]

We remind the reader that the *heat capacity* of a body is defined to be the ratio of the amount of heat[3] absorbed by the body to the resulting rise in its temperature, or in other words the amount of heat energy needed to raise its temperature by one degree.

Note that the heat capacity C_V of a body, say a quantity of gas, held at constant volume V, is less than its heat capacity C_p at constant pressure p. For, if the pressure is held constant, only part of the heat ΔQ supplied to the gas is used to increase its internal energy[4] ΔE, as some is needed to do the work involved in expanding the gas, that is, in increasing its volume by an amount ΔV. Thus we have

$$\Delta Q = C_p \Delta T = \Delta E + p\Delta V, \tag{1.1}$$

where ΔT is the increase in temperature of the gas, and $p\Delta V$ the work done in expanding the gas against the fixed pressure p.[5] Hence

$$C_p = \frac{\Delta E}{\Delta T} + p\frac{\Delta V}{\Delta T}.$$

On the other hand, at constant volume $\Delta V = 0$, we have by definition

$$C_V = \frac{\Delta E}{\Delta T}.$$

[2] Benjamin Thompson (Count Rumford). "Heat is a form of motion. An experiment in boring cannon." *Philosophical Transactions* 88, 1798.

[3] *Heat* is defined as the energy transferred from one body or system to another due to a temperature difference. The *thermal energy* of a body is the total kinetic energy of its atoms and molecules. *Trans.*

[4] The *internal energy* of a body or thermodynamic system is the total of the kinetic energy of its molecules (translational, rotational, and vibrational) and the potential energy associated with the vibrational and electrical energy of the atoms of its molecules. *Trans.*

[5] Since pressure is force per unit surface area of the (expansible) chamber containing the gas, this is indeed the work done in expanding by an amount ΔV. *Trans.*

Hence

$$C_p = C_V + p\frac{\Delta V}{\Delta T}.$$

For a mole of an ideal gas the Clapeyron-Mendeleev state equation[6] holds:

$$pV = RT, \tag{1.2}$$

where $R \approx 8.31$ joules/mole per degree[7] is the ideal-gas constant, and T the absolute temperature (see below). It follows that at constant pressure

$$p\Delta V = R\Delta T,$$

whence the above equation becomes

$$C_p = C_V + R. \tag{1.3}$$

In spite of Rumford's experiments, the caloric hypothesis continued to dominate. There were attempts to combine both points of view; for instance, it was held that the caloric provided the means for the propagation of molecular vibrations and rotations in a body—an exceedingly artificial construction.

We are often baffled by what we learn of the science of the past. For instance, it is difficult to comprehend why belief in the caloric persisted when phlogiston had already been rejected. Chemists of the 18th century postulated phlogiston as a substance constituting the essence of combustibility; that for instance a metal is a compound of ashes ("slag" or "lime") and phlogiston, and when it is burned, the phlogiston escapes and only the ashes remain. The work of Lomonosov and Lavoisier demonstrated that in fact phlogiston was a fiction; combustion of a metal is the same as its oxidation, a synthesis of the metal and oxygen. Thus phlogiston was a sort of anti-oxygen, so to speak.

There were two reasons for the persistence of the caloric hypothesis. Firstly, there were at the time difficulties in explaining the heat of radiation. How might this arise from molecular rotations or vibrations? Secondly, the kinetic theory as it stood at that time offered only qualitative explanations, and was therefore in this respect not at all superior to the caloric theory. The kinetic theory was to triumph at a later stage.

The irreversibility of heat processes

We have already mentioned the dissimilarity of heat and mechanical phenomena. This dissimilarity in fact comes down to the irreversibility of heat processes as

[6] In English texts usually given as deriving from a combination of Boyle's law (1660) and Charles' law (1787). *Trans.*

[7] A *joule* is the work done by a force of one newton (the force needed to accelerate a mass of one kilogram by one metre per second per second) in moving an object through one metre in the direction of the force. A *mole* of a substance contains the same number of particles as 12 grams of carbon-12 atoms, namely 6.022×10^{23}, known as "Avogadro's number". *Trans.*

opposed to those of ordinary mechanics. We shall now consider this theme briefly, leaving the detailed treatment for later.

All phenomena describable in terms of Newtonian mechanics, that is, purely mechanical ones, are reversible. What does this mean exactly? It means that the laws of mechanics do not change if we change the sign of the time, that is, if we replace t by $-t$ everywhere in the formulae of mechanics. This can be seen as follows: Newton's second law states that the force on a body is equal to its mass times the acceleration.

And what is acceleration? The average acceleration of a body over a given time interval from a fixed time t to $t+\Delta t$ is the change in velocity[8] per unit time, over that time interval:

$$\frac{\Delta v}{\Delta t}. \tag{1.4}$$

Since velocity is the change in displacement per unit time, the average velocity over a time interval from t_1 to $t_1 + \Delta t_1$ is

$$v = \frac{\Delta s}{\Delta t_1}.$$

If we consider this average for smaller and smaller Δt_1, then in the limit as Δt_1 goes to zero, we obtain—according to the differential calculus—the velocity of the body at the instant t_1, denoted by

$$v = \left.\frac{ds}{dt}\right|_{t_1} \tag{1.5}$$

to remind us of its origins as the limiting value of the ratio $\frac{\Delta s}{\Delta t_1}$. Hence the average acceleration over the interval from t to $t + \Delta t$ is

$$\frac{\Delta(ds/dt)}{\Delta t},$$

where $\Delta(ds/dt) = \Delta v$ is the change in velocity between times t and $t+\Delta t$. In the limit as Δt goes to zero, this yields the acceleration at the instant t:

$$a = \frac{d(ds/dt)}{dt} = \frac{d^2s}{dt^2}, \tag{1.6}$$

in the notation of the differential calculus.

Observe that in the expression for the acceleration—and hence also the force—the time change enters to the second power. This is clear also from the dimensions of acceleration, namely length per unit time squared. Thus it is that changing the sign of the time does not change the overall behavior of an ideal mechanical system. Mechanical processes are time-reversible.

[8] Here and below displacement and hence velocity and acceleration are vector quantities. Or, for simplicity, it may be assumed that the motion takes place along a straight line in the direction of the force. *Trans.*

But what do we mean by the "sign of the time"? Well, time flows from the past, through the present, and into the future. There appears to be no going backwards in time. A time machine that travels in the negative time direction is impossible.[9]

We normally order events in time so that causes precede effects. This ordering would seem to be secondary to the underlying directionality of time. But such matters are puzzling. We shall not attempt to provide answers to such complex questions, lying as they do on the boundary between physics and philosophy.

Here we desire only to stress the fundamental nature of the difference between the reversible phenomena of classical mechanics and irreversible thermal phenomena. A good way of demonstrating this difference is to play a movie backwards. The processes then depicted in the film that seem compatible with the laws of physics, that is, seem more or less possible, are just the reversible ones, while those that contravene our sense of the possible are the irreversible ones. Scenes of mechanical processes involving the mere displacement of an object retain their naturality when the film is run backwards. Thus if a man is seen to jump off a chair, then in the reverse direction he is seen to jump up onto the chair with his back to it. Although the second action is actually more difficult, it is nonetheless feasible, and does not, therefore, contravene our sense of what is possible. On the other hand, processes such as the burning of a cigarette, the stirring of cream into coffee, and the hatching of an egg, strike us as highly improbable when viewed in reverse.

Reality is irreversible. Hence a mechanics that does not include the concept of irreversibility cannot suffice for understanding the world. In the sequel we shall link the fact that time has a specific direction to this observed irreversibility of certain processes.

A person is born, grows, reaches maturity, grows old, and dies. Life is undeniably irreversible. You may object that these are biological phenomena unrelated to physics. However the fact of the matter is that they *are* related to physics; we will take up this issue below. For the moment let's consider a swinging pendulum. Sooner or later air resistance and friction at the point of suspension will bring it to a halt, a process producing heat, and therefore irreversible since the heat energy produced is not reconverted into the oscillatory energy of the pendulum. To express this mathematically, one must introduce the force due to friction into the equation of motion of the pendulum. Experience shows that in general the force of friction is proportional to a component of the velocity of the body on which it acts, so that the velocity $v = ds/dt$ enters into the equation of motion to the first power, and the invariance under substitution of $-t$ for t disappears.

If two bodies of different temperatures are in contact, then heat (caloric?) flows from the warmer to the cooler body until their temperatures become equal. Thus the past is represented by a temperature difference and the future by equality of temperature. The process of heat flow from one body to another is irreversible.

[9]Or, at least seems to be so? (See Chapter 4.) *Trans.*

In particular it is not possible for a body at uniform temperature to spontaneously divide itself into a hot part and a cool part.[10]

Prior to the discovery of heat engines there were other machines in use, the earliest being the lever and pulley. These machines transformed one form of motion into another, by transferring energy from one source to another. It seemed that nothing was lost in such processes[11], while on the other hand coal or wood in the fire box of a steam engine burn up irreversibly.

In fact a correction is called for here. In our talk of "mechanical phenomena" and of "mechanical engines", we neglected friction, air resistance, and so on. However friction occurs in every mechanical device, causing heat to be produced, and irreversibly transforming mechanical energy into heat.

The difference between ideally mechanical phenomena and heat phenomena turned out to be related to the deepest questions of science. The elucidation of the nature and meaning of irreversible processes became one of the central problems of 19th-century physics—and not only physics.

Since for an ideal (frictionless) mechanical system the sign of the time is immaterial, such a system does not operate irreversibly. However a system involving heat processes is capable of irreversible development, that is, of evolving. Although a clear appreciation of this capacity emerged only 30 years after the appearance of Carnot's memoir, that work had prepared the ground.

It is noteworthy that these same decades witnessed the discovery of the theory of evolution in biology, which found its fullest expression in Charles Darwin's *The origin of species*, published in 1859. Here the subject was evolution in nature, that is, the irreversible development of living things. The connection between biology and physics, which seemed initially of little consequence, subsequently played an enormous part in the development of science, and, surprisingly, in this connection biology had more to impart to physics than vice versa. We shall discuss this in the sequel.

What did Carnot do?

Carnot laid the foundations of "thermodynamics", now one of the main areas of physics.

Thermodynamics is concerned with the general laws that determine the mutual connections between the physical quantities characterizing all those processes occurring in nature and technology, by means of which energy is transferred from one body to another and transformed from one form to another. Thermal processes are included among these. Thermodynamics is a "phenomenological" science, meaning that its concerns are universal and not dependent on the specific

---

[10]Or just highly improbable? (See Chapter 4.) *Trans.*

[11]For example, that the energy used to move one end of a lever is fully available to move the other. *Trans.*

substances involved in the relevant processes of energy exchange. Thus in this respect thermodynamics contrasts with molecular and atomic physics, whose task is to investigate the concrete properties and structure of specific material bodies.

Carnot used the notion of the caloric. In accordance with the ideas of the time, he regarded the caloric as weightless and indestructible. A very strange substance indeed! He wrote: "The motive power in steam engines arises not from an actual loss of caloric, but from its transference from a hot body to a cool one.... In order for motive power to emerge, it is not enough to generate heat: one must also procure coldness; without it heat would be ineffectual...".

Of course, Carnot's idea of the indestructibility of heat—whatever its nature might be—is wrong. However the rest of what we have quoted is correct: for a motive force to emerge, capable of doing work, a temperature difference is indispensable.

Carnot goes on to consider a reversible, cyclical, process: "If we wish to produce motive power by means of the transference of a definite amount of heat from a body A to a body B, we may proceed as follows:

1. Take caloric from body A to generate steam...we assume that steam is formed at the temperature of body A.

2. Funnel the steam into an expansible vessel, for example a cylinder with a piston; the volume of the vessel will then increase along with the volume of steam. The temperature of the steam will decrease as it expands.... We assume that this rarefaction continues until the temperature of the steam reaches that of the body B.

3. Condense the steam by getting it into contact with body B and simultaneously applying constant pressure till such time as it reverts to the liquid state...."

Thus the body A is the heater and body B the cooler. Carnot claims that these three steps can also be carried out in the reverse order: "We can generate steam using the caloric of body B at its temperature, by compressing that body until it heats up to the original temperature of body A, and then condensing the steam by contact with the latter body."[12]

However here the cycle is reversible only in the sense that the system can be returned to its original state in the same way. In fact the transfer of heat from the heater A to the cooler B is really irreversible since external work has to be done to return the system to its original state.

By appealing to the impossibility of a perpetual-motion machine (*perpetuum mobile*), by then already established as a law of nature, Carnot proves that the above-described process yields maximal motive power, that is, work per unit time. His reasoning is remarkable:

"...If there were more efficient means for utilizing heat than the method we have used, that is, if it were possible to obtain a larger amount of power ... then

[12] Presumably following the original process, after which body A has cooled. *Trans.*

one could use a portion of this power to return the caloric by the indicated method from the body B back to the body A, ... and the original state would be restored; one could then repeat the operation and go on in like manner indefinitely: this would not only constitute perpetual motion, but also the unlimited generation of power without the consumption of caloric or any other agent."

Finally, he states his most important conclusion: "The power obtained from heat is independent of the agents used for generating it; ultimately, the amount of power generated is determined exclusively by the temperatures of the bodies participating in the transfer of the caloric."

In other words, the "coefficient of effective action", or "efficiency" of a heat engine is determined by the temperatures of heater and cooler alone, regardless of the particular gas expanding inside it. The greater the difference between these temperatures, the greater the output of work, or "motive power". However Carnot was unable to obtain a quantitative expression for this coefficient, being prevented from doing so by his assumption that caloric is conserved.

Carnot's work is remarkable for the breadth and universality of the analysis, the systematic working out of his thought experiment concerning the cyclicity of the process, and his general conclusions, valid independently of the nature of the particular materials involved.

Carnot's paper went essentially unremarked for a number of years; his contemporaries failed to understand or appreciate it. This has occurred several times in the history of science when ingenious works appeared ahead of their time. Such, for instance, was the case with Gregor Mendel's discovery of the laws of genetics; his work remained unappreciated for 40 years. However today, as a result of the ever increasing internationalization of science and the widespread propagation of information about scientists' work, such situations have become very rare.

The Carnot cycle

In 1834 the French physicist and engineer Benoit-Pierre Clapeyron became interested in Carnot's paper and was able to give it mathematical form—the very form in which the "Carnot cycle" is expounded in modern physics. Such a reformulation was crucial, since no better way has been found of explaining the functioning of a heat engine or introducing the physical concept of entropy into science.

Thus was the indestructible caloric eliminated from the Carnot cycle. But what replaced caloric? The answer is: energy, which will be discussed in the next chapter. Looking ahead a little, we shall see that every body, in particular a quantity of steam or other gas, is characterized under given conditions as possessing a definite amount of internal energy. And what might the phrase "under given conditions" mean for a quantity of gas? Answer: at prescribed temperature and volume.

Thus a body's internal energy is a function of its state, which can change as a result of its interaction with other bodies, more specifically, through being

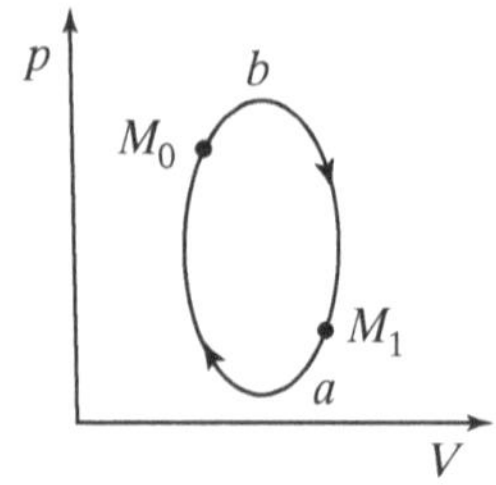

Figure 1.1: States of a gas and transition paths between them.

heated or cooled, or work being done. Thus the change $E_1 - E_0$ in the internal energy[13] of a given quantity of gas is given by

$$\Delta E = E_1 - E_0 = Q - W, \tag{1.7}$$

where Q is the quantity of heat transferred to the gas from the heater, and W is the work done by the gas in expanding.

In equation (1.7) (see also (1.1)) we see that the heat energy Q and work W have equal status.[14] Now heat energy is measured in calories or joules[15], while work is measured in kilogram-metres. Hence it is clear that equation (1.7), which expresses the law of conservation of energy when transformed from one form to another, makes sense only if there is a mechanical equivalent of heat energy, or, conversely, a heat equivalent of work, that is, a conversion factor for converting units of heat into units of mechanical energy, and conversely.

What was the flaw in the theory of the caloric? Answer: It was a mistake to assume that it is possible to determine the amount of caloric, that is, "essence of heat", contained in a body. We shall now deduce from equation (1.7) that the statement "the amount of heat contained in a given quantity of gas is Q" is meaningless.

In Figure 1.1 the states of a fixed quantity of gas are represented by points coordinatized by the pressure p and volume V. The internal energy of the gas in the state M_0 is greater than its internal energy in the state M_1.[16] Suppose that in these two states the gas contains different amounts of heat. Let Q_0 be the amount of heat contained in the gas in state M_0, and

$$Q_1 = Q_0 + Q_{01}, \tag{1.8}$$

the amount when the gas is in state M_1; thus Q_{01} is the change in the amount of heat involved in the transition from state M_0 to M_1, assumed independent of the

[13] See an earlier footnote for the definition.

[14] That is, measure the same sort of entity. *Trans.*

[15] A *calorie* is the amount of heat energy required to raise the temperature of one gram of water at atmospheric pressure by one degree Celsius. One calorie $\approx$ 4.1868 joules. See an earlier footnote for the definition of a joule. *Trans.*

[16] This can be seen, for example, from the description of the Carnot cycle below. *Trans.*

manner in which the transition is effected. According to (1.7),

$$Q_{01} = E_1 - E_0 + W_{01},$$

where W_{01} is the work done by the gas in the course of the transition from state M_0 to M_1.

For a prescribed difference $E_1 - E_0$ in internal energies, we can effect the transition at relatively high pressures, along a path such as b, in order that $W_{01} > E_0 - E_1$, whence $Q_{01} > 0$, which one might interpret as meaning that "the gas contains more heat Q_1 in state M_1 than in state M_0".

However if we effect the transition in the opposite direction, via the path a along which the pressure is low enough for the reverse inequality $W_{01} < E_0 - E_1$ to hold, then we shall have $Q_{01} < 0$, yielding the interpretation "the gas contains less heat Q_1 in state M_1 than in state M_0".

Thus we have arrived at a contradiction. We conclude that "the amount Q of heat in a quantity of gas" depends on the transition path from state to state. Saying that a body contains so much heat is as meaningless as saying that it contains so much work. Unlike the internal energy of a body, heat and work are not functions of the state of the body, but characterize the manner in which energy is transferred from one body to another.

Having liberated ourselves from the caloric (though without for the time being considering the atomic-molecular essence of heat and temperature) we now expound the Carnot cycle.

We have a heater and a cooler. Our working substance is a gas, or more specifically, an ideal gas—a gas rare enough for the interactions between its molecules to be neglected.

The Carnot cycle involves four steps:

a) The gas expands at constant temperature, that is, undergoes a so-called *isothermal* expansion.[17] Since the behavior of an ideal gas is described by the Clapeyron-Mendeleev equation (1.2), namely

$$pV = RT,$$

this isothermal expansion is represented in terms of the coordinates p, V by the hyperbola $pV = RT_1$, where T_1 denotes the temperature of the heater. (See Figure 1.2.)

[17] Achieved, for example, by having the gas expand slowly. *Trans.*

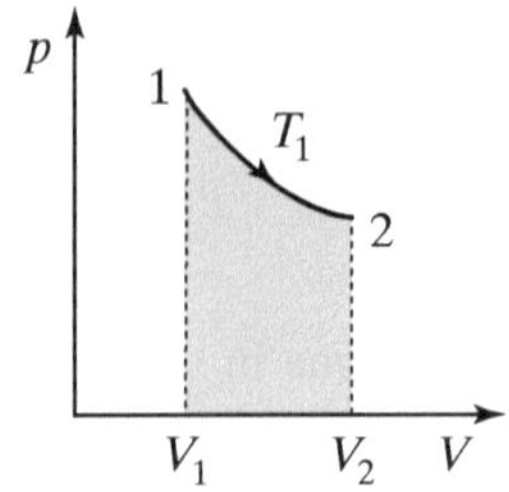

Figure 1.2: Isothermal expansion of the gas.

In so expanding, the gas does an amount of work numerically equal to the shaded area in the figure. This can be seen as follows. In the mechanics of rigid bodies, work is defined as the product of force by distance. In the physics of fluids, since pressure is force per unit area of the surface of the region containing the gas, the force f exerted by the gas in expanding against the piston moving in the gas-filled cylinder is given by $f = ps$, where s is the area of the piston. Hence the differential of the work is

$$dW = f\ dl = ps\ dl = p\ dV,$$

where dl is the distance, that is, an infinitesimal displacement of the piston, and $dV = s\ dl$ an infinitesimal change in the volume of the gas. The work done is then the integral of $dW = p\ dV$ from the initial volume V_1 to the final volume V_2:

$$W = \int_{V_1}^{V_2} p\ dV,$$

which is indeed the shaded area in Figure 1.2.

b) We now insulate the gas from the heater, so that heat is neither supplied to the gas nor withdrawn from it. The gas continues to expand; such expansion without heat exchange is called *adiabatic*. An adiabatic curve is steeper than an isothermal curve, since it is given by an equation of the form

$$pV^\gamma = \text{const.} \tag{1.9}$$

with $\gamma > 1$. (The significance of γ will be explained below.) Adiabatic expansion is accompanied by a lowering of the temperature. We allow the gas to continue expanding to the point 3 lying on the isothermal curve where the volume is V_3 and the gas is at the cooler temperature T_2 (Figure 1.3).

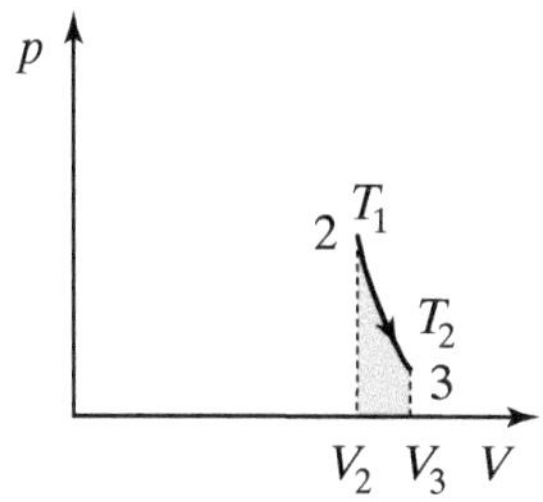

Figure 1.3: Adiabatic expansion of the gas.

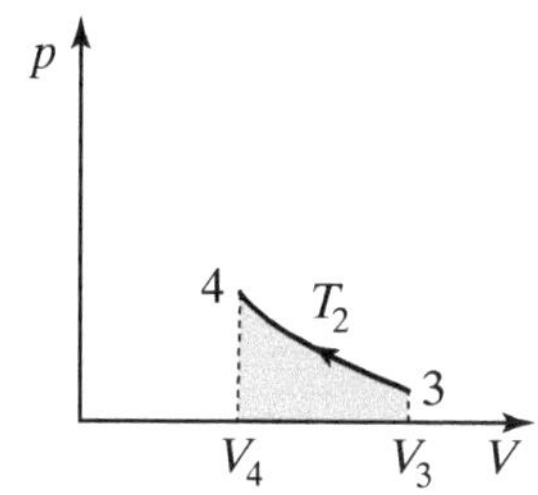

Figure 1.4: Isothermal compression of the gas.

c) At the third stage we compress the gas isothermally at the temperature T_2 by bringing it into contact with the cooler. This requires an amount of work equal to the cross-hatched area in Figure 1.4. We continue compressing the gas till it reaches a volume V_4 such that the corresponding point 4 on the isothermal curve in Figure 1.4 is also on the adiabatic curve through the initial point 1 (Figures 1.2, 1.5).

d) Finally, we again insulate the gas adiabatically and compress it further to its original volume V_1 along the adiabatic curve through the point 1. This requires work to be done (see Figure 1.5).

The cycle is now complete. Positive work has been done in an amount equal to the cross-hatched area enclosed by the cycle (Figure 1.6), that is, the difference between the work done by the expanding gas in stages a) and b) and the work required to compress it in stages c) and d).

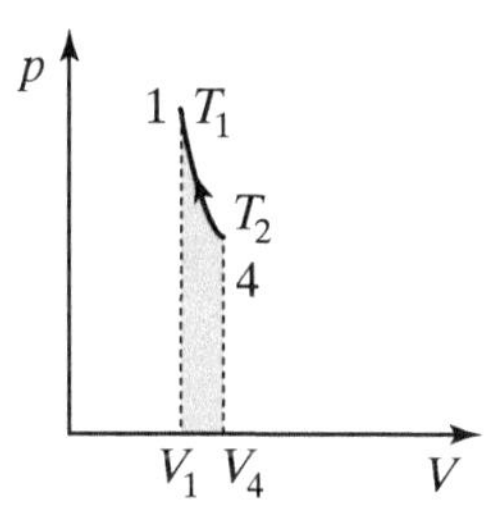

Figure 1.5: Adiabatic compression of the gas.

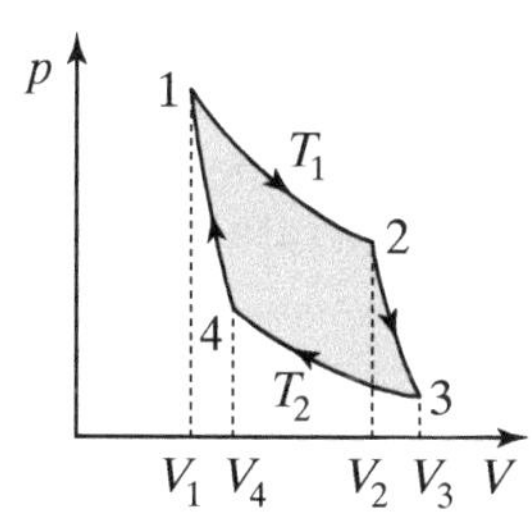

Figure 1.6: The Carnot cycle.

In the course of this process, the gas absorbed from the heater an amount Q_1 of heat at the temperature T_1, and gave out a smaller amount Q_2 of heat at the temperature T_2. Hence the portion of the heat transformed into work is

$$W = Q_1 - Q_2, \tag{1.10}$$

whence

$$Q_1 = W + Q_2 > W. \tag{1.10a}$$

The *efficiency* η of the cycle is then the ratio of the work made available to the heat input:

$$\eta = \frac{W}{Q_1} = \frac{Q_1 - Q_2}{Q_1}. \tag{1.11}$$

How does the efficiency depend on the temperatures T_1 and T_2 of the heater and cooler? We know that Carnot considered this dependence decisive.

Thermodynamic temperature

But what sort of temperature are we talking about here? The temperature in equation (1.2) is the temperature on the Kelvin scale, measured in degrees Kelvin ($°K$), which are equal to degrees Celsius, but start from $-273.15°C$; that is $0°K \equiv -273.15°C$, so-called "absolute zero". How does this "thermodynamic" temperature arise?

Every ideal gas satisfies the state equation

$$pV = \Phi(\theta), \tag{1.12}$$

where Φ is a universal function of the temperature θ measured in degrees Celsius. Experiment shows that this function increases linearly with the temperature, whence it follows that there are constants α, β such that[18]

$$\theta = \alpha pV + \beta. \tag{1.13}$$

The Celsius scale is defined by taking the temperature at which ice melts at atmospheric pressure to be $0°C$ and that at which water boils to be $100°C$. Substituting these two temperatures in the equation (1.13) for an ideal gas, we obtain

$$0 = \alpha(pV)_0 + \beta, \quad 100 = \alpha(pV)_{100} + \beta. \tag{1.14}$$

Solving for α and β from these two equations, we obtain

$$\alpha = \frac{100}{(pV)_{100} - (pV)_0}, \quad \beta = -\frac{100(pV)_0}{(pV)_{100} - (pV)_0}. \tag{1.15}$$

Substituting in equation (1.13) from (1.15), we obtain

$$\theta = \frac{pV - (pV)_0}{(pV)_{100} - (pV)_0} 100°C. \tag{1.16}$$

[18] Since pV is a linear (or affine) function of θ, θ must likewise be such a function of pV. *Trans.*

Experiment shows that for a sufficiently rarefied—that is, ideal—gas

$$\frac{(pV)_{100}}{(pV)_0} = 1.366. \tag{1.17}$$

Equations (1.16) and (1.17) now give

$$pV = (pV)_0(1 + 0.00366\,\theta). \tag{1.18}$$

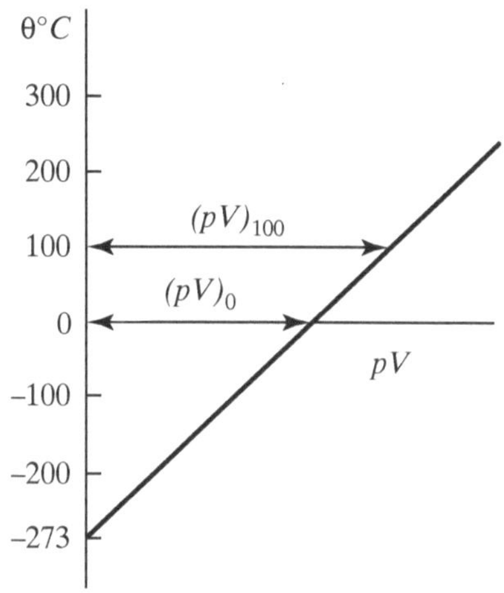

Figure 1.7: Dependence of the "ideal gas temperature" θ on pV.

Figure 1.7 shows the graph of the *ideal gas temperature* $\theta(pV)$ as the function of pV given by (1.18); pV takes on the value zero at $\theta = \theta_0$, given by

$$\theta_0 = -1/0.00366 = -273.15(°C). \tag{1.19}$$

Thus if we take $-273.15°C$ as the zero temperature of the Kelvin scale, then we obtain

$$T°K = \theta - \theta_0 = (\theta + 273.15)°K \tag{1.20}$$

for the temperature measured in degrees Kelvin. At $T = 0°K$, pV vanishes. Substitution from (1.20) in (1.18) yields

$$pV = \frac{(pV)_{0°C}}{273.15}T. \tag{1.21}$$

At $0°C$ and a pressure of 1 atm (atmosphere), the pressure multiplied by the volume taken up by one mole of an ideal gas is known to be

$$(pV)_{0°C} = 22.414\ \ell \cdot \text{atm/mole}. \tag{1.22}$$

(This means that at $0°C$ and 1 atm, a quantity of gas of mass x grams where the molecular weight of the gas is x, occupies a volume of 22.414 liters.) Dividing this by 273.15 yields

$$R = \frac{(pV)_{0°C}}{273.15} = \frac{22.414\ \ell \cdot \text{atm/mole}}{273.15°K} = 8.314 \cdot 10^7 \text{ergs}/(\text{mole} \cdot °K) = 8.314 \text{ joules}/(\text{mole} \cdot °K).$$

We now express the efficiency of the Carnot cycle in terms of T_1 and T_2, the temperatures of heater and cooler.

The amount of heat supplied by a heater at temperature T to the working substance of such a cycle is a function of that temperature:

$$Q = \varphi(T). \tag{1.23}$$

This is a positive function of T which in fact turns out to increase proportionally to the temperature:

$$T = \xi Q. \tag{1.24}$$

It is immediate from this and the second expression for the efficiency in (1.11), that the latter is given by

$$\eta = \frac{T_1 - T_2}{T_1}. \tag{1.25}$$

Thus, in accordance with Carnot's deep ideas, a heat engine with an ideal gas as working substance is characterized as having the maximal coefficient of useful action (efficiency) given by the formula (1.25). Observe that the efficiency is 1 only for $T_2 = 0°K = -273.15°C$.

Finally, we derive a formula for the constant ξ in equation (1.24), using once again the convention that the difference in temperature between melting ice and boiling water be taken as 100 degrees. Thus

$$100 = \xi(Q_{100} - Q_0), \text{ whence } \xi = \frac{100}{Q_{100} - Q_0} > 0. \tag{1.26}$$

Hence equation (1.24) becomes

$$T = \frac{100}{Q_{100} - Q_0} Q \quad \text{or} \quad Q = \frac{Q_{100} - Q_0}{100} T. \tag{1.27}$$

Chapter 2

The laws of thermodynamics

"Everything is possible in this world
Except for prohibitions."
So sing dull poets,
Strumming their lyres.

But nature's stern laws
Are more inspiring than such songs.
The world, truly wonderful,
Obeys those laws unswervingly.

Lomonosov and the conservation laws

The law of conservation (under transformation) of energy also goes under the somewhat old-fashioned and pompous name "the first law of thermodynamics". The second and third laws will soon be revealed. For now we consider just the first.

Who discovered the law? Does it really matter? Why is priority in discovery of any importance?

We live in a real world, not a utopia. In the real world there exist such concepts as national consciousness. It is natural and reasonable for a country, nation, or people to take pride in the achievements of its creative members—its writers, artists, and scholars.[1] Hence a society's interest in who did what first is normal and appropriate—provided such priority is established with the strictest accuracy, and without any hint of the chauvinism that seeks to belittle the achievements of other nations and peoples.

In order to exalt Russian science it has sometimes been claimed that M.V. Lomonosov[2] was the discoverer of the law of conservation of energy. This is

[1] And to feel shame at the misdeeds of its villains? *Trans.*

[2] Mikhail Vasilyevich Lomonosov (1711–1765), Russian polymath, scientist, and writer.

false. Lomonosov discovered the law of conservation of mass[3], founded the kinetic theory of heat, and made many other important contributions to science and the humanities. Pushkin wrote that "He was himself our first university". But the fact remains that he did *not* discover the law of conservation of energy. The great Russian scientist, poet, and artist does not need imaginary discoveries foisted on him!

What is this rumour—which has appeared in print more than once—based on? Answer: On a single sentence in a letter Lomonosov wrote to Leonhard Euler on July 5, 1748. Having described his discovery of the law of conservation of mass or matter, Lomonosov goes on as follows: "Thus the amount of matter gained by a body is lost by another... Since this is a universal law of nature, it extends also to the principles of motion: a body that impinges on another loses as much of its own motion as it imparts to the one it sets in motion".

So much for the conservation of motion. But what does he mean by motion here? The kinetic energy (or *vis viva* as it was called) $mv^2/2$ (m = mass, v = velocity) or the momentum mv (also called "quantity of motion")? A precise notion of energy did not exist in the 18th century, and indeed could not have been formulated until the 19th, the century of steam and electricity.

Lomonosov wrote about the conservation of motion as if it were self-evident, or well known. This is not surprising, since a hundred years earlier, the French philosopher, mathematician, and physicist René Descartes had written: "I claim that there is a known amount of motion in all created matter which never increases or decreases". In his *Principles of philosophy* of 1644 he formulated the "laws of nature", the third of which asserts that "if a moving body encounters another more powerful body, it loses none of its motion; if it encounters a weaker one that it can cause to move, then it loses the amount it imparts to the latter".

Lomonosov knew the works of Descartes intimately and they impressed him more than the rigorous assertions and formulae of Newton's *Principia mathematica.*

Be all this as it may, what *is* true is that Lomonosov discovered the important law of conservation of mass. The idea of conservation laws—the impossibility of getting something from nothing—came to the fore in the physics of the 18th century. (Incidentally, the idea of the "indestructibility" of the fictitious substance called the caloric is related to this development.) In 1775 the French Academy of Sciences announced its refusal henceforth to consider any and all projects having to do with perpetual motion machines.

Such laws represent extremely important general principles pertaining to all of physics. They show that the universe is constructed in a definite way, that nature is governed by certain objective laws. The task of science is to discover these laws, and not at all to seek to refute them.[4] Sometimes one sees laws of nature

[3]This discovery is often attributed to the French chemist Antoine Lavoisier (1743–1794). However, it is clear that Lomonosov has priority. *Trans.*

[4]Some philosophers of science claim that the scientific process consists precisely in attempts to refute such laws, so that our confidence in the correctness of the laws is continually sustained

interpreted as interdictions from on high: Nature forbids perpetual motion, and that's that!

Others unfamiliar with science adopt a different stance. When told that their assertions contradict established scientific laws, they say: "O yes, you say 'such a thing can never be!' but then what seemed impossible yesterday is often realized today!". Further debate is futile.

From time to time, the Biophysics Institute of the Academy of Sciences of the USSR used to produce splendid popular films on scientific subjects. One of the best of these—if not *the* best—showed the first-rate achievements of members of that institute. It bore the striking title "Never say 'never'," implying that everything is possible, there are no impenetrable barriers, no fortresses that cannot be taken by scientists, and so on. However, this is not so. The laws of thermodynamics are forever, just as twice two will always be four.[5]

The law of conservation of energy

How was this law discovered?

We said earlier that 19th century biology did more for physics than physics for biology. This has precisely to do with the first law of thermodynamics.

In 1840, a young German doctor by the name of Robert Mayer found himself in the tropics, on the island of Java. He noticed that the venous blood of people living there was close in color to that of arterial blood—red rather than brown. Mayer knew that the difference in colour of venous and arterial blood is connected with the absorption of oxygen—oxygenated arterial blood is normally a brighter red than deoxygenated venous blood. Body heat results from oxidation, a process akin to burning. Mayer wrote: "The maintenance of the human body at a steady temperature requires that its production of heat be in some quantitative relation to its heat loss, and hence also to the temperature of the surroundings; therefore the production of heat and so *the color difference in the two kinds of blood* must, on the whole, be less intense in tropical latitudes than in cooler countries".

That is how it all began. Mayer arrived at the following general conclusion: "In all physical and chemical processes the force present remains constant". What Mayer called force we now call energy. He wrote further: "The locomotive pulling its train may be compared to a distilling apparatus: the heat furnished to the boiler is transformed into motion, which in turn leaves a residue in the form of heat in the wheel axles". But Mayer went beyond generalities. Denoting the heat capacity[6] of a given specified quantity of air at constant pressure p by C_p and

---

by the fact that they survive such attempts! *Trans.*

[5]Since Einstein we know that energy is *not* conserved, but can be converted to mass and vice versa. Moreover physicists have considered the possibility that the laws of nature might mutate with time or be different in the farthest reaches of the universe. That $2 \times 2 = 4$ would seem to be a truth of a different sort—unless everybody who has done this multiplication has in every case erred, which event has a non-zero, though small, probability. *Trans.*

[6]Defined in Chapter 1.

its heat capacity at constant volume V by C_V, Mayer formulated a quantitative definition of the mechanical equivalent of heat—the conversion factor of heat to work. To this end he set the difference of the heat capacities $C_p - C_V$ equal to the work done by the air in expanding at pressure p (1841). His first measurements (later improved) gave a kilocalorie as equivalent to 365 kilogram-meters.

Thus was the law of conservation of energy discovered.

In 1843, unaware of Mayer's work, James Joule determined the mechanical equivalent of heat by a direct experiment subsequently described in every textbook. Joule heated water in a calorimeter[7] by means of friction—using a little paddle-wheel—and determined the ratio of the work done to the heat generated. He found that a kilocalorie was equivalent to 460 kilogram-meters.

The exact modern value of the mechanical equivalent of heat is 427 kilogram-meters, or $4.18605 \cdot 10^{10}$ ergs $= 4186$ joules.

In 1847 the German scientist Hermann Helmholtz formulated the law of conservation of energy (when transformed from one form to another) in general and rigorous mathematical form. In particular, he proved that energy is given by an integral of motion of the equations of mechanics. It is remarkable that Helmholtz, like Mayer trained to be a doctor, arrived at the law via biological phenomena. He wrote: "According to Stahl, the forces operating in a living organism are physical and chemical forces arising in the organs and substances [of which they are made], but some life force or soul inherent in the organism can arrest or release their functioning... I concluded that Stahl's theory attributes to every living creature the properties of a so-called *perpetuum mobile*.... This suggested the following question to me: What relations must exist between the different forces of nature if one assumes that the *perpetuum mobile* is impossible...?"

Thus by abandoning vitalism in biology, Helmholtz was led to one of the most profound discoveries in physics.

It would seem, therefore, that the first law of thermodynamics, or the law of conservation of energy, was discovered by Mayer, Joule, and Helmholtz in the period 1841–1847. Only much later did it become clear that priority in discovery of the first law belongs to Carnot. It was only in 1878 in connection with a new edition of his *Réflections*, that his notes, hitherto unpublished, appeared in print. There he had written: "Heat is but motive force, or, more correctly, motion that has changed its form; it is the motion of the particles of bodies; wherever motive force is annihilated there comes into being simultaneously an amount of heat exactly proportional to the amount of motive force missing. Conversely, whenever heat disappears, motive force arises.

"We may therefore state the following general principle: The amount of motive force in nature is unchanging. Properly speaking, it is never created and never

[7] A calorimeter is a device used for measuring the heat capacity of a substance, as well as the heat produced in chemical reactions and physical changes. A simple calorimeter consists of just a thermometer attached to an insulated container. From the temperature change, the change in heat is calculated as mass × specific heat × temperature change, where the specific heat of the substance is its heat capacity per unit mass. *Trans.*

destroyed; in reality it [merely] changes form, that is, assumes one or another form of motion, but never vanishes.

"From certain of the ideas I have formed concerning the theory of heat, it follows that the production of a unit of motive force requires the use of 2.70 units of heat."

The figure for the mechanical equivalent of heat found by Carnot (by means unknown) is equivalent to 370 kilogram-meters, which is very close to Mayer's estimate.

Thus Carnot had much earlier abandoned the caloric and given a precise account of the first law. Unfortunately this remarkable work long remained unpublished and unknown.

In our description of a cycle concluding Chapter 1, we tacitly assumed one mathematical version of the first law in using the formula (1.7):

$$\Delta E = Q - W.$$

There we described the Carnot cycle in the modern form due to Clausius,[8] starting with the formula (1.7) and ending with the expression (1.25) for the efficiency of the cycle.

The second law

Carnot's published article contained the second law of thermodynamics. What was said on this theme in Chapter 1 may be compressed into the following assertion:

There is no reversible cyclical process involving the conversion of heat into work that is not accompanied by the transfer of a certain amount of heat from a hotter to a cooler body.

We stress once more that the word "reversible" is used here in a different sense from the one used in mechanics. Earlier we spoke of the reversibility of ideal mechanical processes in time, in the sense that the laws of mechanics continue to hold if such a process is run backwards in the direction of the past instead of the future. In ideal mechanics one can run the film backwards. Here, on the other hand, we are talking only of returning a thermodynamical system to its initial state by means similar to those used in the direct process.

We saw in Chapter 1 that the Carnot cycle is reversible in this sense, and furthermore that it has maximal efficiency compared with any other cycle involving a temperature drop of $T_1 - T_2$.

The reversibility—that is, the closedness of the cycle—comes down to the condition that the isothermally expanding gas is for the whole of that stage of the cycle in thermodynamic equilibrium with the heater, maintaining its temperature at T_1, while during the stage of isothermal compression it is in thermodynamic equilibrium with the cooler, and has its temperature kept at the value T_2. To

[8]Rudolf Julius Emmanuel Clausius (1822–1888), German physicist and mathematician. *Trans.*

ensure this, it is assumed that this expansion and compression take place very slowly—slowly enough for the equilibrium between gas and heat source to remain undisturbed. Such processes are called "quasi-static" since time does not enter into the relations characterizing them.

There is a puzzle here. Irreversibility with respect to time is built into the physics of heat phenomena from the start—heat does not flow on its own accord from a cooler to a hotter body—yet time does not enter into Carnot's fundamental laws of thermodynamics.

Moving on, we observe that the law formulated at the beginning of this section implies the impossibility of a perpetual motion machine "of the second kind". What does this mean?

Clearly the law of conservation of energy would not be contradicted by the transfer of heat from a cooler to a hotter body, or by a situation where work was done as a result of cooling a single heat reservoir, that is, by utilizing heat from a heat source in the absence of a cooler. It does not exclude the unlikely event that a sealed container of water submerged in a bucket of water could come to a boil and the water in the bucket freeze! In each case the amount of energy, whether expressed in calories or joules, remains unaltered. Nor do considerations of energy rule out the possibility of extracting virtually unlimited amounts of useful work from the cooling of the world's oceans. Since the oceans' temperatures are higher than 0°C, or 273.15°K, the amount of heat energy they contain is huge. It is such hypothetical procedures for obtaining work that are called perpetual motion machines *of the second kind.* The first law does not prohibit them, but nonetheless we know that they also are impossible—this amounts to a restatement of the second law of thermodynamics:

A process whose only outcome is the conversion into work of heat extracted from some source is impossible.

This is equivalent to the law as Carnot formulated it; we repeat his version:

A heat engine that absorbs an amount of heat Q_1 at temperature T_1 and yields an amount of heat Q_2 at temperature T_2, cannot do more work than a reversible heat engine, whose work-yield is

$$W = Q_1 - Q_2 = Q_1 \frac{T_1 - T_2}{T_1}. \tag{2.1}$$

This, the second law, is a law of nature. Although as strict and universal as the first, it is of a radically different character.

The pressure of light

The second law provides the key to the solution of a great many problems of physics, chemistry, technology, and, as we shall see, of biology. By way of example we consider a purely thermodynamical proof that light exerts pressure. This proof is due to the Italian physicist Adolfo Bartoli, who proposed it in 1876.

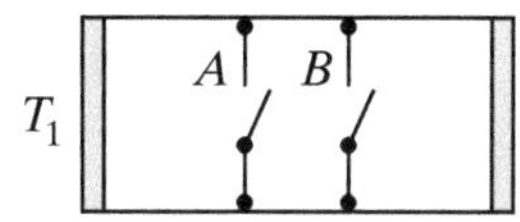

Figure 2.1: Container in Bartoli's proof of light pressure.

We have a container, two of whose opposite walls are at temperatures T_1 and T_2 with $T_1 < T_2$. Movable partitions A and B with openable doors, divide the container into three sections (Figure 2.1). The light given off by each wall fills the corresponding section of the container. The radiation emitted by a wall is in equilibrium with the wall at its temperature. (Bartoli thought of radiation as a sort of gas at some definite temperature, of density proportional to that temperature.) Consider the radiation contained between the wall at temperature T_2 and the partition B. We open the door in this partition, and the radiation fills the two sections bounded on the left by A. We now close the door of B and compress the radiation between the partitions A and B by moving B to the left. By so compressing the "light gas" we raise its temperature to a value exceeding T_1. Next we open the door of A and allow the radiation to reach equilibrium with the left wall, whose temperature will then exceed T_1. In this way we have transmitted heat from the colder wall to the warmer. However, according to the second law this cannot be achieved without some work being done. Work must therefore have been done in compressing the "light gas". This proves that the "light gas", that is, radiation, exerts pressure, which in this experiment resisted the displacement of the partition B.

That light exerts pressure is implied by the electromagnetic theory of light. In 1873 James Clerk Maxwell made this theoretical inference, and was even able to calculate the magnitude of light pressure. In 1901, P.N. Lebedev carried out subtle and precise experiments fully confirming the theory that had enabled Maxwell to compute the pressure of light.

Entropy

We at last come to the concept to which this book is devoted. We rewrite the second equation of (2.1) in the form

$$Q_1/T_1 = Q_2/T_2. \tag{2.2}$$

The quantity Q/T is called *entropy*. It was first introduced by Rudolf Clausius in 1865. The ancient-Greek root of the word "entropy" is $\tau\rho o\pi\eta$, meaning conversion or turning. The verb ἐντρὲπειν means to transform. Hence according to Clausius, entropy is the characteristic feature of transformation, of mutability.

The discussion of Bartoli's thought experiment distracted us from the cyclicity, or reversibility, of thermodynamical processes. Equation (2.2) relates to just

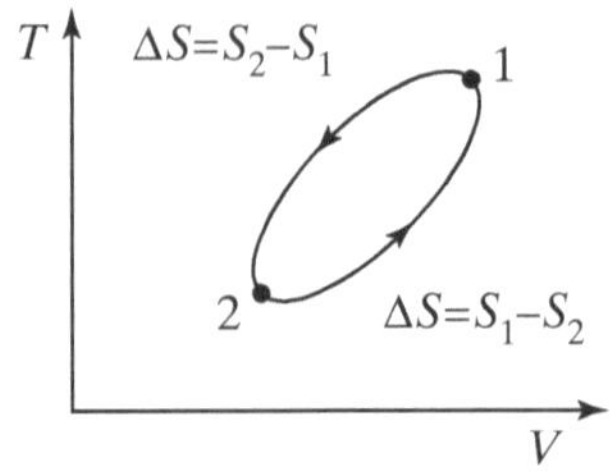

Figure 2.2: A reversible cycle.

such processes. It states that in cyclical, or reversible, processes the amount of entropy absorbed equals the amount exuded, so that in a reversible process the change in entropy is zero. We shall denote entropy by S. Since it is unchanged in a reversible cycle, it is a function of the state of the substance in question: S_1 and S_2 have definite values independently of the way we reached the points 1 and 2 (Figure 2.2). In this entropy resembles internal energy E, but differs crucially from heat Q.

We saw earlier that the heat given off or absorbed is not a function of the state of a system, so that one cannot speak of the quantity of heat contained in a body. Surprisingly, however, if one divides heat by temperature one obtains a function of the state: entropy.

But these words convey only a vague picture of the concept. We need to look at just how entropy is measured.

Clausius argued as follows: If a gas at temperature T absorbs an amount of heat ΔQ, then its entropy is increased by the amount

$$\Delta S = \frac{\Delta Q}{T}, \tag{2.3}$$

or, in terms of differentials,

$$dS = \frac{dQ}{T}. \tag{2.4}$$

To calculate the total change in entropy of the gas in going from state 1 to state 2 (see Figure 2.2) we must integrate the expression (2.4) along each of the indicated paths. However since entropy is a function of the state of the gas, the change in entropy is independent of the path, that is, is determined solely by the initial and final points, whence

$$\Delta S = \int_1^2 \frac{dQ}{T} = S(V_2, T_2) - S(V_1, T_1). \tag{2.5}$$

We stress yet once more that this is not the case for the heat Q; the expression

$$\int_1^2 dQ$$

is indeterminate—the value of the integral depends on the path of integration. In this sense the differential dQ (and likewise the differential of work) is not a "total differential".[9] However on dividing by T the non-total differential dQ turns into a total differential dQ/T—the quantity $1/T$ plays the role of an "integrating factor".

Let us now study the properties of entropy. First, its dimension: Obviously it is measured in units of energy divided by temperature, that is, in ergs/$^\circ K$ or joules/$^\circ K$.

What is the entropy of an ideal gas? During the first stage of a Carnot cycle, the gas does work since it expands isothermally at temperature T_1. This work is done at the expense of an amount of heat Q_1 absorbed from the heater, given by (see p. 14)

$$W_1 = \int_{V_1}^{V_2} pdV = Q_1. \tag{2.6}$$

We shall assume we have one mole of the gas, so that equation 1.2 holds:

$$pV = RT.$$

Since $T = T_1$ is constant during this isothermic stage of the cycle, on replacing p in the integral by RT_1/V, we obtain

$$W_1 = Q_1 = RT_1 \int_{V_1}^{V_2} \frac{dV}{V}. \tag{2.7}$$

Thus we need to consider the integral

$$\int_{V_1}^{V_2} \frac{dV}{V}.$$

We shall see in the next section how to evaluate it. Observe that we effectively specified an isothermal as the path of integration in substituting RT_1/V for p above.

The logarithm and exponential functions

Thus the function to be considered is

$$\int_{V_1}^{V_2} \frac{dV}{V} = \int_{V_0}^{V_2} \frac{dV}{V} - \int_{V_0}^{V_1} \frac{dV}{V},$$

[9] The *total differential* of a smooth function $f(x, y)$ of two variables is the expression $df = \frac{\partial f}{\partial x}dx + \frac{\partial f}{\partial y}dy$, indicating that at any point (x, y) any small changes Δx in x and Δy in y, cause a change in f given to the first order by $\frac{\partial f}{\partial x}\Delta x + \frac{\partial f}{\partial y}\Delta y$. This does not apply to Q since it is not a function of the state variables of the gas (V and T in Figure 2.2). *Trans.*

where V_0 is any value of the volume of the gas less than both V_1 and V_2. We use V_0 as a fixed lower reference point for volume, relative to which we can measure volume in dimensionless units, that is, by replacing V_1 and V_2 by

$$V_1' = \frac{V_1}{V_0} \text{ and } V_2' = \frac{V_2}{V_0}.$$

In terms of such a measure of volume the least lower limit of integration will be 1.

Set

$$f(x) = \int_1^x \frac{dz}{z}. \tag{2.8}$$

We shall show that $f(x)$ has the properties of a logarithm.

Observe first that $f(x)$ is defined only for $x > 0$ (since the integrand $1/z$ is unbounded on intervals including 0) and $f(x) < 0$ for $0 < x < 1$, $f(1) = 0$, and $f(x) > 0$ for $1 < x < \infty$.

Next, consider the function $f(w) = f(ax)$ where a is any positive real number. We have

$$\frac{df(w)}{dx} = \frac{df(w)}{dw}\frac{dw}{dx}.$$

By the Fundamental Theorem of Calculus (asserting that integration can be done by anti-differentiating the integrand) we have $df(w)/dw = 1/w = 1/ax$, whence

$$\frac{df(w)}{dx} = \frac{1}{ax}a = \frac{1}{x}.$$

We see that the derivatives of $f(x)$ and $f(w)$ with respect to x are the same, both equal to $1/x$. This implies that the two functions must differ by a constant C:

$$f(ax) = f(x) + C.$$

In particular we have $f(a \cdot 1) = f(1) + C = C$, whence

$$f(ax) = f(x) + f(a).$$

Thus

$$0 = f(1) = f(\frac{1}{x}x) = f(x) + f(\frac{1}{x}),$$

whence $f(1/x) = -f(x)$.

We see that indeed the function $f(x)$ defined by (2.8) has the properties of a logarithm function. The shape of its graph is shown in Figure 2.3. We write

$$f(x) = \ln x,$$

in standard notation for the so-called *natural logarithm function.*

We now calculate the base of this logarithm, that is, the number e such that for all x, $e^{\ln x} = x$. We saw above that the derivative of $\ln x$ is $1/x$, that is,

$$u = \frac{1}{x} = \lim_{\Delta x \to 0} \frac{\ln(x + \Delta x) - \ln(x)}{\Delta x}.$$

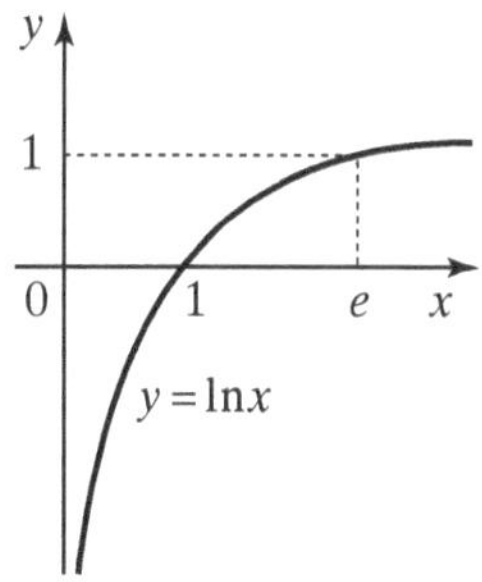

Figure 2.3: Graph of the natural logarithm function.

Setting $\Delta x = i/n$ and letting n tend to infinity, we have

$$\begin{aligned} u &= \lim_{n\to\infty} \frac{\ln(x+1/n) - \ln x}{1/n} = \lim_{n\to\infty} n \ln\left(\frac{x+1/n}{x}\right) \\ &= \lim_{n\to\infty} \ln\left[(1+\frac{1}{nx})^n\right] = \ln\left[\lim_{n\to\infty}(1+1/nx)^n\right]. \end{aligned}$$

Denoting the base of the logarithm ln by e, we infer that

$$e^u = \lim_{n\to\infty}\left(1+\frac{u}{n}\right)^n.$$

Putting $u = 1$ yields

$$e = \lim_{n\to\infty}\left(1+\frac{1}{n}\right)^n = 2.7182818\ldots. \tag{2.9}$$

This is the famous number e, the base of the natural logarithm.

Changing from this logarithm to any other with base b say, is a simple matter. The statement $z = \log_b x$ means that $x = b^z$. Taking the natural logarithm of the latter equation yields $\ln x = \ln b^z = z \ln b$, whence

$$\log_b x = z = \frac{\ln x}{\ln b},$$

or

$$\ln x = \ln b \cdot \log_b x.$$

In particular, taking $b = 10$, we obtain

$$\log_{10} x \approx 0.434 \ln x, \text{ or } \ln x \approx 2.303 \log_{10} x.$$

The exponential function $y = e^x$ has a remarkable property: its derivative is the same as the function itself. To see this, note first that

$$\frac{dy}{dx} = \frac{1}{dx/dy}.$$

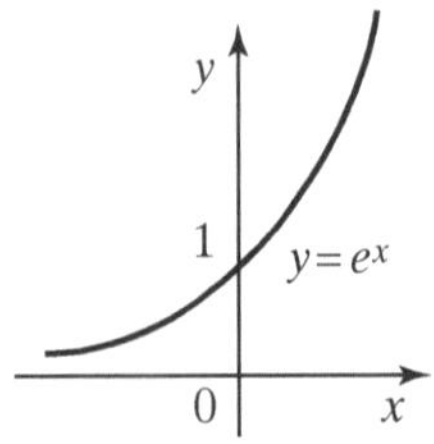

Figure 2.4: Graph of the exponential function.

Now the inverse function is

$$x = \ln y,$$

so that

$$\frac{dx}{dy} = \frac{1}{y}.$$

Hence

$$\frac{dy}{dx} = \frac{1}{1/y} = y = e^x.$$

The graph of the exponential function is shown in Figure 2.4.

Calculation of entropy

Thus according to (2.7),

$$W_1 = Q_1 = RT_1 \int_{V_1}^{V_2} \frac{dV}{V} = RT_1 \ln \frac{V_2}{V_1}. \tag{2.11}$$

Hence the change in entropy due to the isothermal expansion of the gas at temperature T_1 from volume V_1 to V_2 resulting from the absorption of an amount of heat Q_1, is given by

$$\Delta S = \frac{Q_1}{T_1} = R \ln \frac{V_1}{V_2}. \tag{2.12}$$

Since V_2 is greater than V_1, we see that the entropy increases during an isothermal expansion of the gas.

What happens to the entropy as a result of the adiabatic expansion from volume V_2 to V_3? On an adiabatic curve heat is neither supplied to the gas nor taken from it. It follows that an adiabatic process is "isoentropic", does not affect the entropy.

We stated the equation for an adiabatic curve in 1.9:

$$pV^\gamma = \text{ const, where } \gamma > 1.$$

We digress to derive this equation.

If no heat is supplied to a gas then, by the law of conservation of energy,

$$\Delta Q = \Delta E + W = 0. \tag{2.13}$$

In other words, the work is done at the expense of some of the internal energy of the gas:

$$W = p\Delta V - -\Delta E.$$

However on p. 5 we saw that

$$\Delta E = C_V \Delta T,$$

and deduced from the state equation of an ideal gas that

$$p = \frac{RT}{V} = \frac{(C_p - C_V)T}{V}.$$

Substituting these expressions for ΔE and p in (2.13), we obtain the relation

$$\frac{\Delta T}{T} + \frac{(C_p - C_V)}{C_V}\frac{\Delta V}{V} = 0,$$

or, in differential form,

$$\frac{dT}{T} + \left(\frac{C_p}{C_V} - 1\right)\frac{dV}{V}. \tag{2.14}$$

This equation is easy to integrate. In the preceding section we saw that the anti-derivative

$$\int dx/x = \ln x + \text{const}.$$

Hence integration (that is, antidifferentiation) of both sides of the identity (2.14) yields

$$\ln T + \left(\frac{C_p}{C_V} - 1\right)\ln V = \text{const}',$$

or

$$\ln\left[TV^{(\frac{C_p}{C_V}-1)}\right] = \text{const}',$$

whence

$$TV^{\frac{C_p}{C_V}-1} = \text{const}''. \tag{2.16}$$

Substituting for T from $T = pV/R$, we obtain finally the desired equation (1.9):

$$pV^{\gamma} = \text{const}, \text{ where } \gamma = \frac{C_p}{C_V} > 1. \tag{2.17}$$

We now return to the Carnot cycle. For the end-points of the adiabatic expansion from V_2 to V_3 we have from (2.16) that

$$T_1 V_2^{\gamma-1} = T_2 V_3^{\gamma-1},$$

and for the adiabatic compression from V_3 to V_4,

$$T_1 V_1^{\gamma-1} = T_2 V_4^{\gamma-1}.$$

Dividing the former equation by the latter, we obtain

$$\frac{V_2}{V_1} = \frac{V_3}{V_4}. \tag{2.18}$$

Analogously to (2.12), one shows that for stage 3, the isothermal compression of the gas from volume V_3 to V_4, there is a decrease in entropy by the amount

$$\Delta S = R \ln \frac{V_4}{V_3}. \tag{2.19}$$

Hence in view of (2.18), the decrease in entropy over the path $3 \to 4$ is exactly compensated by its increase over the path $1 \to 2$. The cycle closes and the state function entropy remains unchanged.

There is one obvious but important property of entropy, namely that the entropy of a homogeneous system in thermal equilibrium increases in proportion to the mass of the system. This is so because during the transition from some initial state to the state in question, the heat absorbed at each stage of the process is proportional to the mass of the system. This means that the entropy of a system is the sum of the entropies of its homogeneous subsystems. For example, if we have an isolated system consisting of a vessel containing water and water vapor in mutual equilibrium, then the entropy of the system is equal to the sum of the entropy of the water and the entropy of the vapor (and also, of course, the entropy of the material of the vessel). Thus entropy is additive.

In this argument we have not taken into consideration the entropy of the interface between water and vapor, nor of that between the water and the walls of the container. The proportion of molecules involved at these interfaces is relatively small and may be neglected.

Above we calculated the change of entropy when the temperature is held constant. What if the temperature varies? We have

$$dS = \frac{dQ}{T} = \frac{dE}{T} + \frac{pdV}{T} = C_V \frac{dT}{T} + R\frac{dV}{V}. \tag{2.20}$$

Integrating once again, we obtain

$$S = R \ln V + C_V \ln T + a, \tag{2.21}$$

where a is the constant of integration; it is only up to an additive constant that total entropy is determined, since we have only changes in entropy, not "absolute" entropy. From (2.20) we infer that transition from a state V_1, T_1 to a state V_2, T_2 results in the change in entropy given by

$$\Delta S = R \ln \frac{V_2}{V_1} + C_V \ln \frac{T_2}{T_1}. \tag{2.22}$$

Hence a rise in temperature is accompanied by a rise in entropy provided that the change is not adiabatic. As we have already seen, along an adiabatic curve entropy does not change—in the case of adiabatic expansion of a gas, the increase in entropy due to the increase in volume is exactly compensated by its decrease due to the resulting cooling of the gas.

The above calculations are subject to certain provisos. For instance we tacitly assumed that C_p and C_V do not vary with the temperature. This is in fact not completely true; in fact the heat capacity of a body increases with decreasing temperature, and this increase is especially marked at low temperatures. Thus the rigorous theory is more complicated than the one just presented.

We next compute the change in entropy resulting from heat conduction equalizing the temperatures of gases, and from diffusion, that is, the mixing of gases.

Consider an adiabatically insulated system consisting of two identical vessels each of volume V and each containing a mole of an ideal gas at temperatures T_1 and T_2. The vessels are brought into contact and as a result of heat-conduction in their walls the gases reach a state of thermal equilibrium without change in volume. According to (2.21) the total entropy prior to the time of contact is

$$S = 2R\ln V + C_p \ln T_1 + C_p \ln T_2 + 2a.$$

After contact the equilibrium temperature of both gases is

$$T = \frac{T_1 + T_2}{2},$$

and the entropy is

$$S' = 2R\ln V + 2C_p \ln\left(\frac{T_1 + T_2}{2}\right) + 2a.$$

The change in entropy is therefore

$$\Delta S = S' - S = C_p\left\{\ln\left[\frac{T_1 + T_2}{2}\right]^2 - \ln(T_1T_2)\right\}.$$

It is easy to see that the entropy has increased: This is immediate from the fact that the arithmetic mean of two positive quantities is always greater than or equal to their geometric mean:

$$\frac{T_1 + T_2}{2} \geq \sqrt{T_1T_2}.$$

Here is a proof. We need to show that (squaring both sides and multiplying by 4)

$$(T_1 + T_2)^2 \geq 4T_1T_2,$$

or, equivalently, that

$$T_1^2 + 2T_1T_2 + T_2^2 \geq 4T_1T_2,$$

that is,

$$T_1^2 - 2T_1T_2 + T_2^2 \geq 0.$$

But this is true since the left-hand side is $(T_1 - T_2)^2$, a square and so never negative.

Thus the entropy has increased by the amount

$$\Delta S = C_p \ln \frac{(T_1 + T_2)^2}{4T_1T_2}. \tag{2.23}$$

We now turn to the mixing of two gases. Suppose the first gas occupies a volume V_1 and contains n_1 moles, and the second contains n_2 moles at volume V_2. We assume the two gases are at the same temperature T and pressure p, and are separated by a partition. The partition is removed and the gases mingle. How does the entropy change from the situation where the gases are separated to that where they are mixed?

Since the volume of each gas in the mixture is $V_1 + V_2$, the individual changes in entropy of the gases are

$$\Delta S_1 = n_1 R \ln \frac{V_1 + V_2}{V_1} \text{ and } \Delta S_2 = n_2 R \ln \frac{V_1 + V_2}{V_2}.$$

The combined change in entropy due to the mixing of the gases, the so-called "mixing entropy", is then the sum:

$$\Delta S = \Delta S_1 + \Delta S_2.$$

Since

$$V_1 = n_1 \frac{RT}{p} \text{ and } V_2 = n_2 \frac{RT}{p},$$

this becomes

$$\Delta S = R \left(n_1 \ln \frac{n_1 + n_2}{n_1} + n_2 \ln \frac{n_1 + n_2}{n_2} \right). \tag{2.24}$$

Thus the mixing entropy ΔS is positive.

We see from this that entropy increases in processes occurring spontaneously, such as those involving heat conduction and diffusion. If such processes could be reversed then entropy would decrease, but such processes—producing a difference in temperature or in concentrations of gases in a mixture—would require external work to be done.

Measuring entropy experimentally

How can entropy be measured in practice? By definition, the entropy change in a substance between 0°K and temperature T_1 is

$$\Delta S = \int_0^{T_1} \frac{dQ}{T} = S_{T_1} - S_0.$$

Now there is a third law of thermodynamics, Nernst's[10] "heat theorem", postulating that the entropy S_0 vanishes at absolute zero.[11] (We shall discuss Nernst's theorem further below.) Hence we may write

$$S = \int_0^{T_1} \frac{dQ}{T}. \tag{2.25}$$

What is the increase dQ in heat? If heating takes place at constant pressure, then

$$dQ = C_p dT, \tag{2.26}$$

whence

$$S = \int_0^{T_1} C_p \frac{dT}{T}. \tag{2.27}$$

However, in having its temperature raised from absolute zero to T_1, the substance in question may be subject to various effects and its heat capacity may well change.

Consider, for instance, carbon tetrachloride $CC\ell_4$. At room temperature, 298.1°K, this compound is a partially vaporized liquid. Now at low temperatures the heat capacity of pure crystalline substances is known to be proportional to the cube of the temperature:

$$C_p = bT^3. \tag{2.28}$$

In the case of $CC\ell_4$ the constant b is approximately $0.75 \cdot 10^{-3}$. At 225.4°K a change takes place in the crystal lattice of solid frozen $CC\ell_4$ requiring 1080.8 calories per mole to effect. At the melting point 250.2°K of $CC\ell_4$, the latent heat of melting, that is, the amount of heat needed to change the crystalline solid into a liquid, is 577.2 calories/mole. At its boiling point of 298.1°K, the latent heat of vaporization is 7742.7 calories/mole. Hence in units of calories/(mole·°K) we have

$$\begin{aligned} S_{298.1} \quad = \quad & \int_0^{10^\circ K} 0.75 \cdot 10^{-3} T^2 dT + \int_{10^\circ K}^{225.4^\circ K} C_p \frac{dT}{T} + \frac{1080.8}{225.4} \\ & + \int_{225.4^\circ K}^{250.2^\circ K} C_p \frac{dT}{T} + \frac{577.2}{250.2} + \int_{250.2^\circ K}^{298.1^\circ K} C_p \frac{dT}{T} + \frac{7742.7}{298.1}. \end{aligned}$$

At 298.1°K the pressure of carbon tetrachloride vapor is 114.5 mm of the mercury column, equivalent to 0.15 atm. To bring the vapor to atmospheric pressure we must compress it, and this entails a decrease in entropy of

$$\Delta S_{298.1} = R \ln \left(\frac{V_{760}}{V_{114.5}} \right).$$

Using values of C_p found by experiment, the integrals with integrand $C_p dT/T$ can be expressed explicitly as functions of the temperature T, and then integrated

[10] Walther Hermann Nernst (1864-1941), Prussian physicist, Nobel laureate in chemistry 1920. *Trans.*

[11] And thus giving meaning to "absolute" entropy S after all. *Trans.*

Table 1: The entropy of carbon tetrachloride at 298°K and 1 atmosphere.

Change in entropy in calories/(mole ·°K)	S cal./(mole·°K)
$S_{10} - S_0 = 0.75 \cdot 10^{-3} \int_0^{10} T^2 dT$	0.25
$S_{225.4} - S_{10}$ (integrated graphically)	36.29
$\Delta S_{225.4} = 1080.8/225.4$ (phase transition)	4.79
$S_{250.2} - S_{225.4}$ (integrated graphically)	3.08
$\Delta S_{250.2} = 577.2/250.2$ (melting)	2.31
$S_{298.1} - S_{250.2}$ (integrated graphically)	5.45
$\Delta S_{298.1} = 7742.7/298.1$ (vaporization)	25.94
$R \ln(114.5/760)$ (compression)	-3.76
Total: $S = 74.35$ cal/(mole·°K) $= 311.22$ joules/(mole·°K)	

Table 2: Values of C_p and C_p/T for silver and graphite at various temperatures.

	C_p in cal/(mole·°K)		C_p/T in cal/(mole·°K^2)	
T°K	silver	graphite	silver	graphite
50	2.69	0.13	0.0531	0.0026
100	4.82	0.41	0.0482	0.0041
150	5.54	0.79	0.0379	0.0053
200	5.84	1.22	0.0292	0.0060
250	5.97	1.65	0.0259	0.0066
273.1	6.02	1.86	0.0221	0.0068
298.1	6.04	2.08	0.0203	0.0069

graphically, that is, by drawing the graph of C_p/T as a function of T, and estimating the area under the graph between the respective limits of integration (compare Figure 2.5 below).

The individual terms making up the entropy $S_{298.1}$ of $CC\ell_4$ at 298.1°K and atmospheric pressure, together with their sum (the value of $S_{298.1}$) are given in Table 1 above.

In this way the entropy of various substances may be estimated experimentally. We shall see that often its value can be calculated theoretically. The air in a room, a sheet of paper, any object whatsoever, contains a definite amount of entropy—just as it contains a definite amount of internal energy.

We now give two further examples where the entropy is calculated by means of graphical integration. Table 2 gives values of C_p and C_p/T for silver and graphite at various temperatures and a certain fixed pressure p. The respective graphs of C_p/T as a function of T are sketched in Figure 2.5.

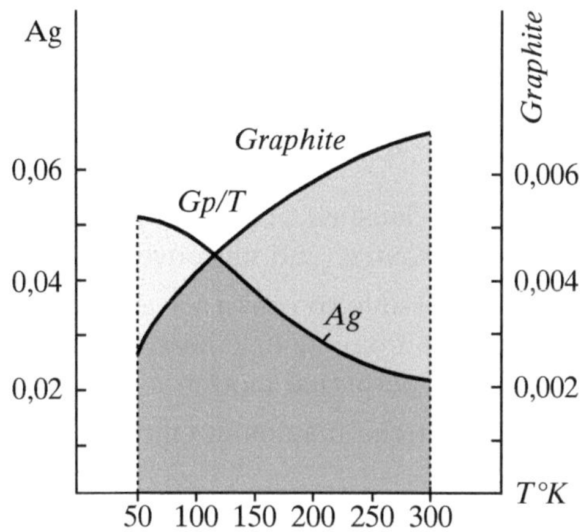

Figure 2.5: Graphical calculation of the entropy of silver and graphite.

One first calculates the approximate average values of C_p/T over various of the indicated intervals. For example, over the interval from 50°K to 100°K the average value $\bar{C}_p/T$, in calories/(mole·°K) for silver is $(0.0531 + 0.0482)/2 = 0.0506$. One then multiplies each of these averages by the length of the corresponding temperature interval, and sums these products to obtain an approximation of the desired integral:

$$\sum_{i=1}^{n} \frac{\bar{C}_{p,i}}{T_i} \Delta T_i \approx \int_{T_1}^{T_n} \frac{C_p dT}{T}.$$

For silver this sum is

$$50(0.0506 + 0.0425 + 0.0330 + 0.265) + 0.0230 \cdot 23.1 + 0.0212 \cdot 25$$
$$= 8.69 \text{ cal/(mole} \cdot^\circ \text{K) or } 36.38 \text{ joules/(mole} \cdot^\circ \text{K)}.$$

For graphite the sum is

$$50(0.0034 + 0.0047 + 0.0056 + 0.0063) + 0.0061 \cdot 23.1 + 0.0068 \cdot 25$$
$$= 1.31 \text{ cal/(mole} \cdot^\circ \text{K) or } 5.48 \text{ joules/(mole} \cdot^\circ \text{K)}.$$

These figures represent the changes in entropy of silver and graphite when heated from 50 to 298.1°K.

Irreversible processes

We state once more the laws of thermodynamics—of which there are now three, rather than two.

The first law (Mayer, Joule, Helmholtz, Carnot). An increase in internal energy of a system is the sum of the heat absorbed by the system and the work done on the system:

$$dE = dQ + dW.$$

The second law (Carnot, Clausius). A process whose *exclusive* outcome is the extraction of heat from a heat source (and its conversion into work) is impossible.

In other words, it is impossible to construct an engine that works cyclically and does work by drawing heat from a single heat reservoir, without causing any other changes in the system (a perpetual motion machine of the second kind).

We saw in Chapter 1 that the maximum efficiency a reversible heat engine can have is

$$\eta = \frac{W}{Q_1} = \frac{T_1 - T_2}{T_1}.$$

Alternative version of the second law (Clausius, Thomson[12]). If in a reversible process a system absorbs an amount ΔQ of heat at temperature T, then the entropy of the system increases by the amount

$$\Delta S = \frac{\Delta Q}{T}.$$

Entropy is a function of the state of the system.

The third law (Nernst, 1906). At $T = 0°\mathrm{K}$, the entropy $S = 0$. (Why this is so will be explained below.)

So far in our study of entropy we have encountered nothing especially interesting, let alone mysterious. However entropy does possess one surprising peculiarity: While—like energy—it is preserved in reversible processes, unlike energy it increases in irreversible ones.

Suppose that we bring two bodies into contact, having temperatures T_1 and T_2 with $T_1 > T_2$—for instance, that we drop an ice cube at temperature T_2 in a glass of water at temperature T_1. The water will then transmit an amount of heat ΔQ to the ice, causing the water's entropy to decrease by $\Delta Q/T_1$, while the heat ΔQ absorbed by the ice cube will increase its entropy by $\Delta Q/T_2$.[13] The overall change in entropy is positive:

$$\Delta S = \frac{\Delta Q}{T_2} - \frac{\Delta Q}{T_1} > 0. \tag{2.29}$$

We have already seen that entropy increases in spontaneous processes such as diffusion and heat conduction. Such processes can be reversed only at the net expense of work. And one might adduce a great many other examples attesting

[12] Joseph John Thomson (1856–1940), British physicist, discoverer of the electron. *Trans.*

[13] This assumes that the temperatures of ice cube and water are not substantially changed. *Trans.*

to the invariable increase of entropy in irreversible processes. Equalization of the temperatures of two bodies can occur reversibly (as in the first two stages of a Carnot cycle; see (2.12)), and irreversibly. In the latter case the increase in entropy will be greater.

At what point does the increase in entropy of a system stop? Answer: It increases until the system reaches a state of equilibrium. Here we come to questions of a fundamentally new sort. In our discussions of reversible processes we avoided considering the flow of such processes in time, assuming that they proceed infinitely slowly and that there is equilibrium at each stage. Along the isothermals of a Carnot cycle, it was assumed that the temperature of the gas was kept throughout the same as that of the heater or cooler respectively. Strictly speaking, we have been dealing with thermostatics rather than thermodynamics, since we ignored time-related features such as the rate of progress of the processes.

By contrast, an irreversible process involves in an essential way a progression in time towards equilibrium. It takes time for the temperatures of a hot and a cold body brought into contact to become equal. Entropy does not attain its maximum instantaneously.

So far (so far!) time has not appeared explicitly in our discussions. Yet we were nonetheless concerned with the dynamics of heat processes. To remedy this lack, we supplement our earlier formulation(s) of the second law with a proviso in the case of irreversible processes:

Addendum to the second law (Clausius, 1865). In irreversible processes the total entropy of the system always increases, that is, [14]

$$dS > \frac{dQ}{T}, \qquad (2.30)$$

or

$$S > \int_0^{T_1} \frac{dQ}{T}. \qquad (2.31).$$

In arriving at this conclusion Clausius and Thomson brought to light a fundamental property of the universe as a whole.

In actuality there are no reversible processes. A tiny portion, at least, of the mechanical energy of every motion is transformed into frictional heat. Sooner or later all moving bodies come to a halt, and an equilibrium corresponding to maximum entropy is reached. Thomson concluded that the world is ultimately subject to "heat death"—while its energy remains unchanged.

There is a further very general conclusion to be drawn. We saw early on that in pure mechanics there are in principle no irreversible processes, while in thermodynamics there are such processes. We have tacitly assumed that, for isolated irreversible systems, time increases in the direction of increase of entropy. We shall

[14] The equations that follow are perhaps intended to indicate that if a system undergoes an irreversible process, then entropy is generated internally, in addition to that resulting from heat input. (See also equation (6.5) below.) *Trans.*

see in the sequel how such matters stand for open systems interacting with the surrounding world with its matter and energy.

Thus the concept of entropy has led us from mere technology (the steam engine) to cosmological considerations (the direction of time and the fate of the universe).

Thus entropy turns out to have remarkable properties. We shall in the sequel search for the reason behind this. However first we turn again to reversible processes and thermostatics, in order to derive some interesting and important consequences of the laws of thermodynamics.

Chapter 3

Entropy and free energy

Energy is the mistress of the world,
But a black shadow
Follows her inexorably,
Making night and day one,
Emptying everything of value,
Transforming all to smoke-filled gloom....
At least that's how entropy
Has been invariably represented.
But now we know
That there is no such shadow
And never was nor will be,
That over the successive generations of stars
There is only entropy—life and light.

Obtaining useful work

As we have seen, the equation giving a system's change in internal energy when undergoing a reversible process is

$$dE = dQ - dW, \tag{3.1}$$

where dQ is the heat absorbed by the system and dW the work done by the system. We also have (see (2.4))

$$dQ = TdS. \tag{3.2}$$

Hence

$$dE = TdS - dW, \qquad (3.3).$$

or

$$dW = -(dE - TdS) = -dF, \tag{3.4}$$

where

$$F = E - TS \tag{3.5}$$

is called the *Helmholtz free energy* of the system. This form of the equation in effect combines the first two laws of thermodynamics, asserting that the work done by the system is not merely at the expense of some of its internal energy but at the expense of the internal energy less the heat. Thus the greatest useful work that a system can do is equal to (the loss of) its free energy.

We wish also to consider processes—reversible or not—proceeding at constant pressure p.[1] In such a process work is done against the constant pressure, regardless of reversibility. As we have seen, this work is given by pdV. The energy remaining for doing other useful work is then

$$dW = -dF - pdV. \tag{3.6}$$

Thus at constant pressure, the amount of energy available for doing useful work[2] is equal to the quantity

$$G = F + pV = E + pV - TS. \tag{3.7}$$

The quantity G is called the *Gibbs*[3] *free energy*, or *thermodynamic potential*, of the system. Like internal energy and entropy, both kinds of free energies are functions of the state of the system, that is, their values depend only on the state of the system,[4] and not on the transition path to that state from some other.

It follows from (3.6) and (3.7) that the maximal amount of useful, *non-expansive* work obtainable from the system at constant pressure is equal to the loss of thermodynamic potential:[5]

$$dW = -dG. \tag{3.8}$$

If the volume is also constant, then equation (3.6) reduces to equation (3.4), pertaining to the case of an arbitrary reversible process: the maximal amount of energy available for useful work is equal to the decrease in Helmholtz free energy:

$$dW = -dF.$$

So far we have talked only of mechanical work, that is, the work pdV done by an expanding gas. However equation (3.8) is valid for arbitrary kinds of work, such

[1] Of particular interest in connection with chemical reactions occurring at atmospheric pressure. *Trans.*

[2] Of a non-mechanical sort. *Trans.*

[3] Josiah Willard Gibbs (1839–1903), American physicist, chemist, and mathematician, one of the originators of vector analysis. *Trans.*

[4] Meaning that they are determined by any two of the "state variables" p, V, T. The fact that they are state functions is thus immediate from their definitions. *Trans.*

[5] The work here is that not involved in expansion, that is, the maximal obtainable work less the work done in expanding against the fixed pressure. *Trans.*

as that done by an electrical current, or as the result of a chemical reaction, and so on. The various kinds of work transform into one another.[6] For instance, the chemical reaction in a battery produces electric current, which in turn can be used to do mechanical work—such as when the current from a car battery activates the starter. The same result can be obtained by using a crank, in which case muscular energy is brought to bear instead. But where does that energy come from?

A muscle is a mechano-chemical system in which free chemical energy is used directly for doing mechanical work.

A heat engine uses heat obtained from the chemical reaction of burning fuel. In accordance with Carnot's law (2.1) some (but not all) of the heat is changed into mechanical work. However a living organism exists at constant temperature and pressure. This means that the muscular work performed by people and other animals—such as that carried out in building the Egyptian and Mexican pyramids—is not to be explained in terms of a transfer of heat from heater to cooler. The work done by muscles transforms chemical free energy produced in specific chemical reactions taking place in muscles at constant temperature.

In what direction do these chemical reactions proceed? We shall now answer this question.

Equilibrium conditions

If a system is isolated (closed and insulated), that is, does not exchange matter or energy with the surrounding world, then spontaneous processes taking place within the system will tend towards equilibrium.[7]

A purely mechanical process evolves without giving off or absorbing heat; it is an adiabatic process involving no change in entropy. Here the role of free energy is played by the total mechanical energy. As is well known, mechanical equilibrium is reached at a minimum of potential energy. A stone tossed upwards falls to the ground. Thus the condition for mechanical equilibrium under the assumption $\Delta S = 0$, is that potential energy be at a minimum:

$$U = U_{\min}, \tag{3.9}$$

and the direction of such a spontaneous process is that of decreasing potential energy:

$$\Delta U < 0. \tag{3.10}$$

Since the total mechanical energy of a system undergoing a frictionless process is constant, any decrease in the potential energy is exactly made up by an increase in the kinetic energy. This transformation can be exploited as a source of

[6] Or work transforms one kind of energy into another, which is then once more available for doing work. *Trans.*

[7] Although there are systems—such as planetary ones—that do not seem ever to settle into any steady state. *Trans.*

useful work—as , for example, in hydroelectric power stations, where the potential energy of the water is turned into kinetic energy as it falls, and then by means of turbines into electric energy.

As noted earlier, in an isolated adiabatic system equilibrium is reached at the greatest value of entropy:

$$S = S_{\text{max}}. \tag{3.11}$$

A good example of an isolated system is a spaceship, since in constructing the ship every effort is made to ensure that its interior is insulated as far as possible from the surrounding cosmos. Of course, the spaceship, with its functioning crew of astronauts, is far from being in a state of equilibrium. Later on we shall see how the non-equilibrium state of a living organism is sustained. After all, for such a system equilibrium means death.

Thus the direction of spontaneous irreversible change in an isolated system is that of increasing entropy:

$$\Delta S > 0. \tag{3.12}$$

For the time being we leave aside the question of the rate of increase of entropy, that is, the speed at which equilibrium is attained.

We turn now to a closed system at constant temperature and pressure. (A system is said to be *closed* if it can exchange energy but not matter with the surrounding medium.) An example would be a chemical reaction taking place in a flask whose temperature is controlled by a thermostat. The equilibrium condition for such a system is the minimality of the Gibbs free energy:

$$G = G_{\text{min}}, \tag{3.13}$$

and consequently the direction of change of state of such a system is that of decreasing free energy:

$$\Delta G < 0. \tag{3.14}$$

We have seen above (see (3.7)) that the Gibbs free energy consists of two parts:

$$G = (E + pV) - TS = H - TS.$$

The internal energy plus pV is a state function called the *enthalpy*[8] of the system:

$$H = E + pV. \tag{3.15}$$

We mention in passing that the term "heat content" sometimes used for enthalpy is unfortunate, since, as we have seen, it makes no sense to speak of the amount of heat in a body.

Thus in a closed system at constant temperature and pressure, we must have

$$\Delta G = \Delta H - T\Delta S \leq 0. \tag{3.14a}$$

---

[8]Enthalpy is simply another convenient state function. For an isobaric (constant pressure) process, we have $\Delta H = \Delta E + p\Delta V$, so from the energy equation $\Delta E = \Delta Q - \Delta W$ we obtain $\Delta H = \Delta Q$, the heat absorbed at constant pressure. *Trans.*

We see that such a decrease in Gibbs free energy can come about in two distinct ways: through a decrease in enthalpy or an increase in entropy. Of course these may occur together, or indeed the enthalpy may increase but have its growth more than compensated by an increase in entropy:

$$\Delta S > \frac{\Delta H}{T} > 0.$$

And, finally, the reverse may occur: enthalpy and entropy might decrease together with the loss of enthalpy exceeding that of entropy. Briefly, for a process of the sort we are considering to be realizable, the Gibbs free energy must decrease; it is not sufficient to consider the changes in enthalpy and entropy separately.

A chemical reaction

The burning of any kind of fuel—an oxidation process—is accompanied by a reduction in its free energy, which is given off in the form of heat and light. Thus the burning of hydrogen—its combination with oxygen—resulting in the formation of water, involves a large discharge of free energy. Specifically:

$$H_2 \text{ (gas at 1 atm) } + \frac{1}{2}O_2 \text{ (gas at 1 atm) } \to H_2O \text{ (liquid) } + 236760 \text{ joules/mole.}$$

On the right-hand side of this equation we see the amount of free energy liberated. Since the reaction takes place at constant pressure, this must be the Gibbs free energy, or thermodynamic potential.

Before it was understood that what counts is free energy rather than internal energy or enthalpy, it was thought that a chemical reaction is possible only if it is exothermic, that is, involves the production of heat. However, in time it was realized that there are endothermal reactions, involving the absorption of heat. The direction of change in enthalpy is not in itself the crucial factor.

Liberation of free energy is a necessary condition for a chemical reaction to take place, but it is not sufficient. For example, a mixture of hydrogen and oxygen can exist for an indefinitely long time without a reaction occurring. However if a lighted match is brought up to the mixture an explosive reaction occurs.

What is going on here? It turns out that, although it furnishes us with the condition (3.13), thermodynamics does not tell us whether or not a reaction will proceed—only whether it can or cannot occur. For example, under ordinary conditions oxygen and nitrogen will not react because

$$N_2 + O_2 \to 2NO - 174556 \text{ joules/mole,}$$

that is, the free energy increases rather than decreases in this reaction. At the same time this indicates the thermodynamic instability of nitric oxide, since the reverse reaction results in the liberation of free energy:

$$2NO \to N_2 + O_2 + 174556 \text{ joules/mole.}$$

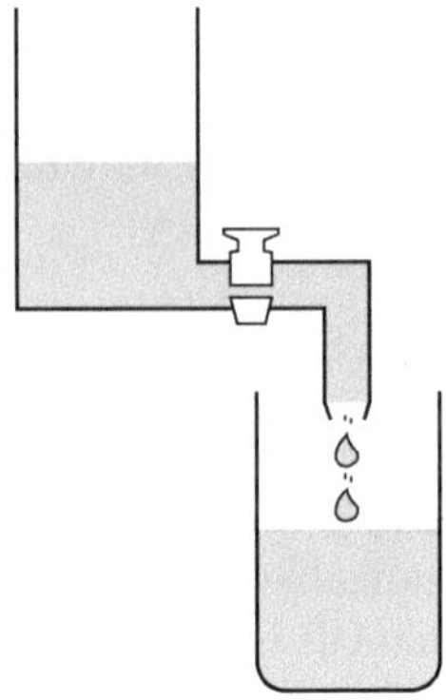

Figure 3.1: Model of a chemical reaction in an isolated system.

So why is it that thermodynamics tells only that a reaction can in principle occur and not whether it will actually take place?

The fact of the matter is that a process may be thermodynamically possible, yet *actually* impossible.[9] It was noted above that, although it is thermodynamically possible for hydrogen and oxygen to react when mixed—since the reaction liberates free energy—, such a mixture may nonetheless remain inert for an arbitrarily long time. Thermodynamics has nothing to say about us igniting the mixture with a match—it merely finds how the balance of free energy lies.

Every process takes time. If the time required is infinite, it will not proceed. However, time does not enter into thermodynamics; as mentioned earlier, strictly speaking we should call it thermostatics rather than thermodynamics.

Imagine a system consisting of a vessel containing a liquid with a tap for draining the liquid into a second vessel placed under the tap (Figure 3.1). Thermodynamics tells us that sooner or later the liquid from the upper vessel will all flow into the lower, and of course we can calculate the final level of liquid in the lower vessel. Thermodynamics tells us that the liquid must flow downwards, not upwards. But will this occur sooner rather than later? Thermodynamics is silent on this question. The rate of flow depends on the extent to which the tap is turned on. If it is turned off, then the liquid in the upper vessel is in a state of thermodynamical disequilibrium since its energy is above the equilibrium value for the system. Nevertheless this disequilibriun can last indefinitely.

There is a clear similarity between igniting a mixture of hydrogen and oxygen with a match and turning on the tap: opening the tap allows the process to proceed.

Figure 3.2 shows the graph of the free energy against time over the course of a certain chemical reaction. The initial state 1 of the reagents has greater free energy than the final state 2 of the products of the reaction:

$$G_1 - G_2 = \Delta G > 0.$$

[9] In the sense that some external input is necessary to set it off? *Trans.*

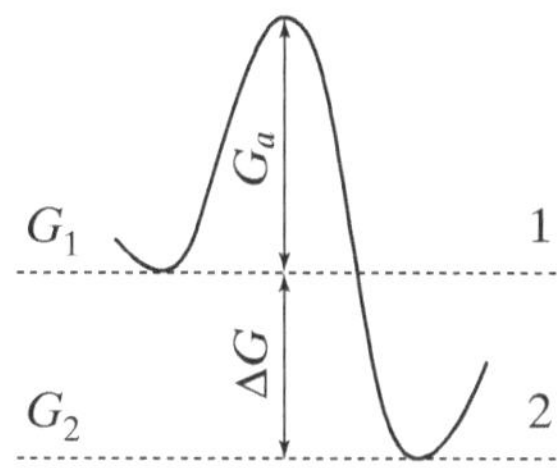

Figure 3.2: Variation in free energy during a chemical reaction.

Free energy has ultimately been liberated. Thus the reaction is possible. But will it occur? This is a question of kinetics rather than thermodynamics. In the reaction in question the initial state is separated from the final state by a so-called "activational barrier", a kind of ridge which must be surmounted by the reagents. The free energy at the top of the ridge exceeds the initial free energy G_1 by the amount G_a. This means that for the reaction to proceed the reacting molecules must possess a surplus of free energy. The higher the barrier, the slower the pace of the reaction. If the temperature is raised, then the proportion of molecules with surplus energy increases and the reaction speeds up. (We shall see later exactly how the speed of a chemical reaction depends on the free activation energy and the temperature T.)

The reader doubtless knows that many reactions are carried out with the aid of catalysts, substances that stimulate the reactions while themselves remaining unchanged. Under conditions of chemical equilibrium, the role of the catalyst reduces to the lowering of the activational barrier and consequent speeding up of the reaction. Consider again the vessels and liquid of Figure 3.1. Here a catalyst might be an agent who "opens the tap" more, thereby accelerating the flow of liquid from the upper to the lower vessel. However the final result is independent of this catalyzing agent: sooner or later the liquid will all flow into the lower vessel and reach a predetermined level in it. (Of course the system depicted in Figure 3.1 is not analogous to a chemical reaction with an activational barrier—the flow of liquid is hampered only by friction. However, it helps to clarify the distinction between thermodynamics and kinetics.)

The activation free energy is equal to the difference between the activation enthalpy and the activation entropy times the temperature (see the equation following (3.14)):

$$G_a = H_a - TS_a. \tag{3.16}$$

Thus a reduction in G_a can be brought about by a reduction in the activation enthalpy H_a, or an increase in the activation entropy S_a, or some other combination of changes.

It is appropriate to mention here that, without exception, all the chemical reactions taking place in living organisms—and which in fact constitute life—

involve catalysts. These are the albumins[10] and enzymes.

Melting of crystals and evaporation of liquids

In a closed system under equilibrium conditions, on being heated crystals melt into a liquid, and on further heating the liquid turns into a vapor, that is, a gas.

The melting of a crystal and the evaporation of a liquid require the input of heat, the *latent heat of fusion* and *the latent heat of evaporation* respectively. If one places ice in a vessel and heats it, when the temperature reaches $0°C$ ($\equiv$ 273.15°K) it stays steady until all of the ice has melted. One can then continue heating the water to $100°C$ ($\equiv$ 373.15°K), whereupon the temperature again ceases increasing until such time as all the water has turned into steam. The latent heat of melting of ice is 5982 joules/mole, and the latent heat of evaporation of water at 100°C is 40613 joules/mole. The latent heat of evaporation of a substance is always significantly greater than its latent heat of melting.

Crystal, liquid, and gas are different *phases* of a substance. They differ in their state and existence conditions, and are separated by boundaries when they coexist. A more technical name covering both melting and evaporation is *phase transition.*

We shall now find thermodynamical conditions for phase transition. When a crystal is heated, its internal energy increases, and hence also its enthalpy H. So too does its entropy—it was shown in Chapter 2 (see equation 2.22) that entropy increases with increasing temperature provided the change is not adiabatic. Nevertheless, as a rule the (Gibbs) free energy $G = H - TS$ also increases. As the temperature continues to rise, the free energy finally becomes equal to the free energy of the same quantity of liquid at the melting point:

$$G_{crystal} = G_{liquid}, \tag{3.17}$$

or

$$H_{crystal} - T_{melting} S_{crystal} = H_{liquid} - T_{melting} S_{liquid}. \tag{3.18}$$

When the free energies of the two phases coincide, a phase transition occurs—melting, in the present case. We can solve for the melting point $T_{melting}$ from (3.18):

$$T_{melting} = \frac{H_{liquid} - H_{crystal}}{S_{liquid} - S_{crystal}} = \frac{\Delta H}{\Delta S}. \tag{3.19}$$

As noted above, both ΔH and ΔS are positive. The greater ΔH—which is in fact the latent heat of fusion—and the smaller the change in entropy ΔS, the higher the melting point $T_{melting}$. If the entropies of crystal and liquid should coincide, so that $\Delta S = 0$, then melting could not occur: $T_{melting} \to \infty$. Thus an entropy change at the melting point is crucial for melting to be possible.

[10] Certain water-soluble proteins responsible for the functioning of cells. *Trans.*

Let's calculate the fusion entropy $\Delta S_{melting}$ for water. The latent heat of fusion for ice is known to be $\Delta H = 5982$ joules/mole, and of course $T_{melting} =$ 273.15°K. Hence

$$\Delta S_{melting} = \frac{5982 \text{ joules/mole}}{273.15°\text{K}} = 21.89 \text{ joules/(mole} \cdot° \text{K)}.$$

(We did the analogous calculation for carbon tetrachloride in Chapter 2—see Table 1 there.)

Vaporization entropy is calculated similarly. For water it is

$$\Delta S_{boiling} = \frac{40613 \text{ joules/mole}}{373.15°\text{K}} = 108.84 \text{ (joules/mole} \cdot° \text{K)}.$$

We see, therefore, how important a role entropy plays. Without changes in entropy there would be no phase transitions, and the world we inhabit could not exist. In particular, if at least there were water, then it could exist only as ice—and there would be no life.

Up till now we have been considering these physical processes merely phenomenologically, or formally. We considered the changes of enthalpy and entropy involved in melting and evaporation, but did not ask "Why?", which for physics is the fundamental question. Why do enthalpy and entropy decrease when a substance is cooled and increase when it is heated? Why are the latent heat and entropy of evaporation so much greater than the latent heat and entropy of fusion?

Once again, thermodynamics is silent on these questions; they go beyond the bounds of its competence. The three laws by themselves can never lead to answers.

Of course physics can answer these questions, but by means of theories developed in other areas of physics, namely statistical mechanics and the kinetic theory of matter. We shall look at these in the next chapter.

Why does alcohol dissolve in water yet gasoline not do so?

Now that we know about chemical reactions and phase transitions, we may, after reflecting a little, also master the thermodynamical theory underlying the dissolving of one substance in another.

The most important and universal solvent is of course water. Water solutions are everywhere in our lives. Actually, tap water is a solution. It contains various dissolved substances, mainly hardening salts—carbonates, silicates, phosphates, predominantly of calcium. These salts are not very soluble, and they therefore gradually settle out on the walls of vessels or water-pipes as a scale. This scale is a nuisance. Though not much of a problem in a teapot, it can cause a great deal of trouble in a steam boiler.

So these salts have low solubility in water. For which substances is it high?

Alcoholic drinks come in many different strengths, which shows that ethyl alcohol dissolves in water over a wide range of proportions of alcohol to water. We often need to dissolve sugar and table salt in water, and everyone knows that their solubility increases with the temperature. These are just the most familiar examples of the great many water-soluble substances—salts, acids, bases—whose solubility, as a rule, increases with the temperature.

However, there are many substances that do not dissolve in water. It hardly needs mentioning that mercury is one such substance—the idea of a solution of mercury in water strikes one as unnatural. (Yet silver *does* dissolve to a very small extent in water, yielding a solution useful as a bactericide.) A great many organic compounds, above all hydrocarbons, are practically insoluble in water. Thus gasoline and paraffin, which are mixtures of hydrocarbons, form layers in water, as is shown, for example, by the iridescent film formed by gasoline on the surface of puddles. As a result of the high surface tension of water, the layer of gasoline is stretched so as to form a thin film whose colors have the same origin as those of soap bubbles. The colors visible in such thin films result from the phenomenon of interference, and so provide a beautiful proof of the wave nature of light.[11]

But what is the thermodynamical significance of greater or lesser solubility?

Clearly, a solution will form if its formation is accompanied by a reduction in free energy, much as in a chemical reaction:

$$\text{soluble substance} + \text{ solvent} \rightarrow \text{ solution.}$$

So free energy must be released—and this can only come about through an increase in entropy.[12] It follows from equation (2.22) that at constant temperature entropy increases with increasing volume, and also that mixing entropy is positive (equation (2.24)). When we mix a liter of alcohol and a liter of water we do not restrict the mixture volume-wise, so the entropy increases. In addition, the enthalpy decreases as a result of the interaction of the molecules of the solute and the solvent. Hence the free energy must decrease:

$$\Delta H < 0 \text{ and } \Delta S > 0, \text{ whence } \Delta G = \Delta H - T\Delta S < 0.$$

That entropy increases when many substances are dissolved is proved by the increase in their solubility with temperature (as in the case of sugar or salt dissolved in water). Since the contribution ΔH of the change in enthalpy turns out to be relatively small, the increase in solubility resulting from a temperature increase

[11] Actually the full explanation comes from quantum electrodynamics, which considers light as made up of photons. See R.P. Feynman's *QED*, Princeton University Press, Princeton, 1985. *Trans.*

[12] Since the enthalpy H decreases (see below) and the temperature T is unchanged when a solution forms. *Trans.*

must be due to the term $-T\Delta S$ having decreased (that is, become larger negatively). Hence the higher the temperature the more free energy available to be liberated—provided ΔS is positive.

Thus alcohol, sugar, and salt dissolve in water because their dissolution is accompanied by a decrease in free energy. But why does gasoline not dissolve in water? The obvious answer is that this would require an increase in free energy.

And why is that? Well, free energy can increase in two distinct ways—either through an increase in enthalpy or a decrease in entropy. Which of these is more pertinent to the case in question?

Hydrocarbons do dissolve in water, but only to a minimal extent. Careful investigation has shown that this process is accompanied by a decrease in enthalpy: $\Delta H < 0$. But what is very unusual is that the solubility of hydrocarbons goes down rather than up as a result of heating. It follows that the entropy must also decrease during dissolution—moreover by an amount sufficient to compensate for the reduction in enthalpy:

$$\Delta H < 0 \text{ and } \Delta S < 0, \text{ but } \Delta G = \Delta H - T\Delta S > 0,$$

that is, the positive quantity $-T\Delta S$ exceeds the loss ΔH of enthalpy: $-T\Delta S > |\Delta H|$.

We conclude that the layering of gasoline and water is due to entropy! The entropy change causes the hydrocarbon molecules to be expelled from the watery environment. The entropy change plays the role of an acting force!

Hydrophobic forces and the albumin globule

This "entropic force" is usually called a *hydrophobic force*, and substances that this force expels from water are also termed *hydrophobic*; they are "inimical" to water, unlike hydrophilic substances which "like" water.

Many important phenomena can be explained by the hydrophobic force. For example, how does soap clean?

Soaps are usually made from sodium and potassium salts of fatty acids, with chemical formulae similar to the following one (for sodium palmitate, a sodium salt of palmitic acid):

$$\begin{array}{l} H_3C{-}CH_2{-}CH_2{-}CH_2{-}CH_2{-}CH_2{-}CH_2{-}CH_2{-}CH_2{-} \\ \qquad {-}CH_2{-}CH_2{-}CH_2{-}CH_2{-}CH_2{-}CH_2{-}COONa. \end{array}$$

The long hydrocarbon group $H_3C(CH_2)_{14}$ is hydrophobic, while the radical COONa is hydrophilic. In water, soaps form colloidal solutions,[13] and the suspended molecules form mycelia,[14] that is, molecules like $HC_3(CH_2)_{14}COONa$ arranged in a specific manner, with the hydrophobic hydrocarbon groups oriented

[13] Very fine suspensions of particles in a liquid. *Trans.*

[14] A *mycelium* is something like a mass of fibers. *Trans.*

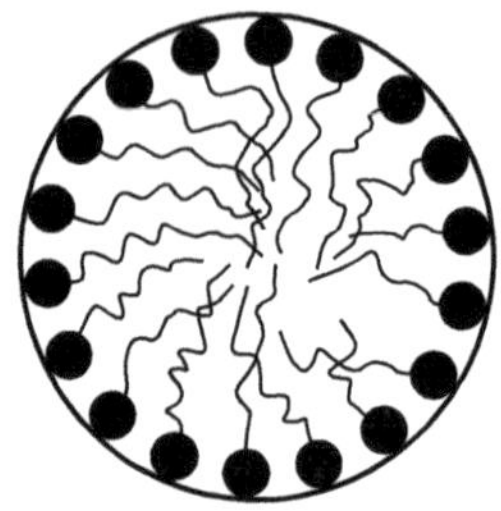

Figure 3.3: Model of a soap mycelium, showing the hydrophilic "heads" and hydrophobic "tails" of molecules.

towards the interior of the mycelium—being repelled by the water—while the hydrophilic groups remain on the surface of the mycelium. See Figure 3.3 for a schematic picture of a soap mycelium.

The cleansing action of soap is due to the mycelial structure of its solution. The surfaces of the mycelia are highly active, and readily adsorb[15] many substances. The presence of the hydrophilic groups enables soap to wet hydrophobic surfaces and to emulsify fats, oils, and so on. This is coupled with the alkaline reaction of the COONa radical.

However the most important effects of hydrophobic entropic forces are not rainbow films on water or the cleansing action of soap, but the synthesis of albumins, substances determining the functioning of all life processes.

An albumin molecule is a chain of amino acid residua. They are all built out of the 20 different aminoacids, according to the following general scheme:

$$\begin{array}{c} \mathrm{H} \\ | \\ \mathrm{H_2N—C—COOH} \\ | \\ \mathrm{R} \end{array}$$

where R stands for a functional group of atoms distinguishing one aminoacid from another. When aminoacids combine to form an albumin chain, water molecules separate off and peptide links —CO—NH— are formed. Here is a fragment of an albumin chain (tripeptide):

$$\begin{array}{ccccccc} & \mathrm{H} & & \mathrm{H} & & \mathrm{H} & \\ & | & & | & & | & \\ \mathrm{—NH—} & \mathrm{C} & \mathrm{—CO—NH—} & \mathrm{C} & \mathrm{—CO—NH—} & \mathrm{C} & \mathrm{—CO—} \\ & | & & | & & | & \\ & \mathrm{R_1} & & \mathrm{R_2} & & \mathrm{R_3} & \end{array}$$

where R_1, R_2, and R_3 stand for different or identical radicals. An entire chain, which may be very long—containing a hundred or more aminoacid residua—is a

[15]That is, adhere to and surround in a thin layer of particles. *Trans.*

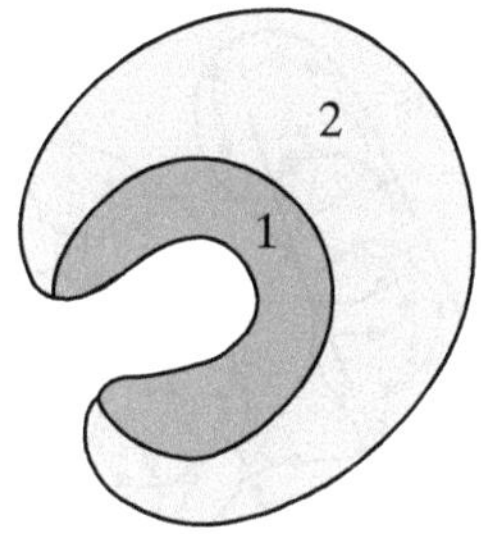

Figure 3.4: Schematic sketch of an albumin globule. Region 1 consists of hydrophobic radicals, and region 2 of hydrophilic ones.

kind of text written using a 20-letter alphabet. We shall have more to say about these texts at the end of the book.

Albumins function in water solutions as fermenting agents, catalysts of biochemical reactions. They possess a certain definite flexibility since rotations are possible about the unit links C—N and C—C. In its naturally-occurring biologically functional state, an albumin chain is rolled up into a dense globular structure, characteristic of that particular albumin. What determines this structure? Among the radicals, or functional groups $R_1, R_2, \ldots, R_{20}$, there are hydrophobic ones, containing hydrocarbon groups, and hydrophilic ones, containing in particular acidic and basic groups. What happens to these different sorts of radicals when an albumin chain is immersed in water?

To some extent the structure of an albumin globule resembles that of a mycelium, in the sense that the albumin chain is rolled up so that the hydrophobic groups, which spontaneously withdraw themselves from the surrounding water, are located in the interior of the globule, while the hydrophilic radicals are located on its surface. Figure 3.4 gives a schematic picture of such a globule, and Figure 3.5 gives a more detailed representation of the structure of a globule of the albumin myoglobin, established by means of X-ray analysis.

Many albumins function naturally in globular form. This naturally-occurring state can be undone by means of acids, alkalis, or the application of heat. The albumin then becomes denatured and ceases to function. You can't get a chicken out of a boiled egg.

We have arrived at a highly non-trivial conclusion: the functioning of several of the albumins crucial to life depends on their globular structure, and this is determined by hydrophobic, that is, entropic, forces acting in a watery environment.

What do rubber and an ideal gas have in common?

This would seem to be a silly question, or poorly framed at best. What could a solid substance—which rubber certainly is—have in common with a gas, moreover

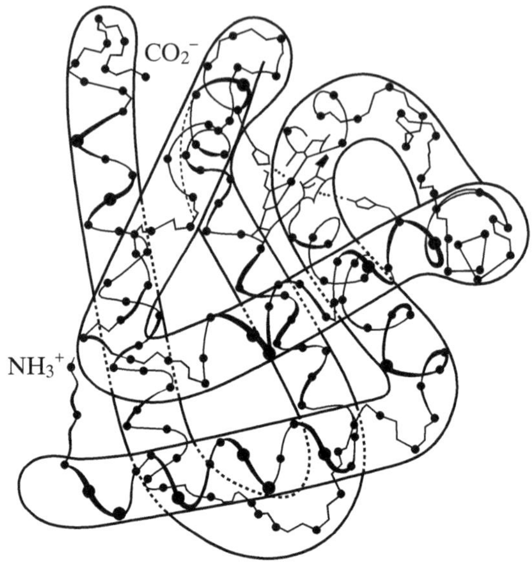

Figure 3.5: Structure of a globule of myoglobin. The dots represent aminoacid radicals. In myoglobin the albumin chain has a roughly spiral structure, as shown here.

a perfect, that is, highly rarefied, gas? It must be, surely, that the poser of the question had in mind some specific property of rubber in which it differs from other solids but resembles an ideal gas.

However, it is precisely such seemingly paradoxical questions that often lend impetus to scientific progress. L.D. Landau[16] used to say that the task of theoretical physics is to establish new connections between phenomena that at first sight have nothing in common. The finding of such connections invariably proves to be a potent source of new insights. (A striking example is Maxwell's discovery of the connection between the wave theory of light and theory of electromagnetism.)

In the present case, our interest lies with elasticity: elastic force and deformation.

When a steel spring is stretched there arises an elastic force, which increases with the amount of stretching. According to Hooke's law, the tension developed in a flexible solid is proportional to the deformation:

$$\sigma = \frac{f}{s} = \epsilon \frac{L - L_0}{L_0}, \tag{3.20}$$

where σ denotes the tension, that is, the force f per unit area of cross-section of the deformed body,[17] L is the length of the body in its stretched state, L_0

[16] Lev Davidovich Landau (1908–1968), Soviet physicist, Nobel laureate 1962. *Trans.*
[17] That is, cross-section perpendicular to the line of stretching. *Trans.*

its length unstretched, and ϵ is Young's modulus of elasticity of the material of the body.[18] For steel the modulus is very large, around 200 gigapascals.[19] Thus a small deformation (or strain) results in a large elastic tensile force, or, conversely, a large applied force (stress) is needed to produce an appreciable deformation of the spring.

Whence comes the elastic force? When we deform a strip of metal elastically we raise its internal energy, by increasing the potential energy of its atoms, held together in the crystal lattice of the metal by chemical bonds. So what changes is just the internal energy. Now we see from equation (3.4) that the work done in deforming the metal is equal to the increase in its (Helmholtz) free energy:

$$\Delta W = \Delta F = \Delta E - T\Delta S.$$

But work is force times distance, so we also have

$$\Delta W = f\Delta L = f(L - L_0). \tag{3.21}$$

Hence the elastic force is given by

$$f = \sigma s = \frac{\Delta F}{\Delta L} = \frac{\Delta E}{\Delta L} - T\frac{\Delta S}{\Delta L}. \tag{3.22}$$

As already noted, when a steel spring is deformed, its internal energy changes but not its entropy.[20] Hence $\Delta S = 0$, and the elastic force is given by $f = \Delta E/\Delta L$, so that it is of a purely energetic character.

Most solids behave this way under elastic deformation—but not rubber. The elastic modulus of rubber is many orders of magnitude less than that of steel; depending on the degree of vulcanization, it ranges from 200 to 8000 kilopascals—which means simply that a piece of rubber, a rubber band for instance, can easily be stretched elastically to several times its length.

We next calculate Young's modulus for an ideal gas. A gas may also be considered an elastic body, but one that resists compression rather than stretching. Suppose our ideal gas is contained in a cylinder with a piston, as in Figure 3.6. When the gas is compressed isothermally there arises an elastic force

$$f = ps = \frac{\Delta F}{\Delta L}.$$

Since here $L = V/s$, where now s is the area of cross-section of the cylinder, it follows that

$$f = ps = s\frac{\Delta F}{\Delta V} = s\left(\frac{\Delta E}{\Delta V} - T\frac{\Delta S}{\Delta V}\right). \tag{3.23}$$

---

[18]Determined by this equation. *Trans.*

[19]Since 1 pascal = 1 newton/meter^2, a gigapascal is 10^9 newtons/m^2 ($\approx 30 \times 10^6$ pounds/square inch). *Trans.*

[20]Since $\Delta Q = 0$? Or is the author maintaining that $\Delta S = 0$ follows from (3.22)? This is unclear in the original. *Trans.*

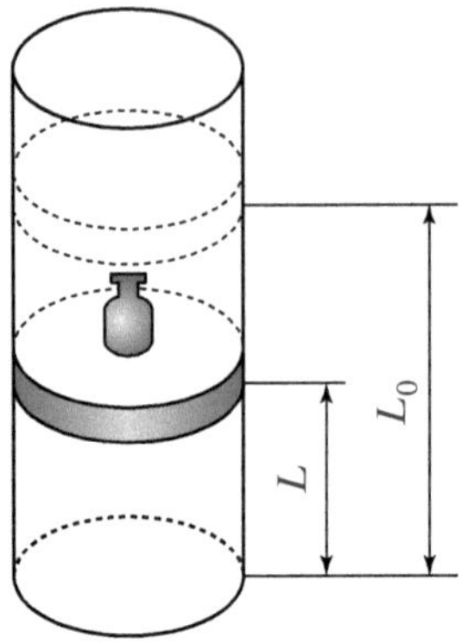

Figure 3.6: Compression of a gas in a cylinder with a piston.

Now the basic assumption concerning an ideal gas is that there be no interaction between its molecules beyond elastic collisions. Hence the internal energy of a quantity of ideal gas is independent of the average distance between the molecules, and hence of the volume. Thus $\Delta E = 0$, and (3.23) yields

$$p = -T\frac{\Delta S}{\Delta V}. \tag{3.24}$$

For an ideal gas we have the equation of state

$$pV = RT.$$

Since the compression is isothermal, T is fixed. Hence

$$pV = \text{const},$$

whence

$$d(pV) = V dp + p dV = 0,$$

yielding

$$dp = -p\frac{dV}{V}. \tag{3.25}$$

This equation is analogous to Hooke's law, if we interpret dp as the "elastic tension" of the gas and dV/V as its relative deformation; the negative of pressure, $-p$, is then the analogue of the elastic modulus.[21] Note that the "elastic modulus" $-p$ of the gas at a particular volume V is proportional to the temperature T, since by the state equation,

$$p = \frac{RT}{V}.$$

[21]Except that the elastic modulus for metals, for example, is close to being constant, whereas here $-p$ is variable. *Trans.*

This may also be seen from equation (3.24).[22]

Equation (3.25) shows that the elasticity of an ideal gas is not energy-based like that of a steel spring, but entropic: a gas resists compression not because compression increases its energy but because it decreases its entropy.

Atmospheric pressure, or the "elastic modulus" of an ideal gas, is around 100 kilopascals, which is of the same order of magnitude as the elastic modulus of rubber. It turns out that the elasticity of rubber is also entropic. Experiment shows that the elastic force f of rubber is proportional to the absolute temperature T, and furthermore is close to zero at absolute zero. Hence in (3.22) only the term $-T(\Delta S/\Delta L)$ is significant, that is,

$$f \approx -T\frac{\Delta S}{\Delta L}. \tag{3.26}$$

Thus the high elasticity of rubber is explained by the fact that its entropy decreases very markedly under stretching.

In view of the fact that the elasticity of an ideal gas and a rubber band are both essentially entropic, one would expect similar heat phenomena to be observable for these two substances. And indeed, anyone who has pumped up a bicycle tire or stretched a rubber band held against his or her lips will attest to the fact that in each case heat is given off. Quick compression of a gas is an adiabatic process because there is no time for heat to be absorbed by the surrounding medium, and the same is true with regard to the rapid stretching of a rubber band.

That the elasticity of rubber is essentially entropic is very important: the main use of rubber, namely in automobile and airplane tyres, depends on this property. We conclude this section by juxtaposing in tabular form the elastic properties of an ideal gas and of rubber (Table 3 below).

However we have still not penetrated to the underlying reason for this similarity of such radically different material bodies. We shall consider this in the sequel.

Why do we heat a room?

This apparently very simple question will allow us to better understand how energy and entropy are interrelated.

The outstanding theoretical physicists Arnold Sommerfeld and Ryogo Kubo both included in their monographs on thermodynamics a note written by the Swiss geophysicist Robert Emden[23] entitled "Why do we have winter heating?",

[22] Young's modulus also decreases with temperature for metals, though relatively slightly. For example, for carbon steel it decreases by about 6% between 0°C and 100°C. *Trans.*

[23] Jacob Robert Emden (1862–1940), Swiss astrophysicist and meteorologist. *Trans.*

Table 3: Properties of an ideal gas compared with those of rubber.

Ideal gas	Rubber
Elastic modulus is proportional to absolute temperature, and equals 100 kilopascals at one atmosphere.	Elastic modulus is proportional to absolute temperature, and lies between 200 and 8000 kilopascals.
Volume can be changed by a large factor.	Length can be changed by a large factor.
Heats up under adiabatic compression.	Heats up under adiabatic stretching.
Internal energy independent of volume.	Internal energy practically independent of length.
Under compression an entropic elastic force arises.	Under stretching an entropic elastic force arises.

published in the British journal *Nature* in 1938. We follow the example of these scientists and quote Emden's note in its entirety.[24]

"The layman will answer [to the question as to why we have winter heating]: 'To make the room warmer.' The student of thermodynamics will perhaps so express it: 'To import the lacking (inner, thermal) energy.' If so, then the layman's answer is right, the scientist's wrong.

"We suppose, to correspond to the actual state of affairs, that the pressure of the air in a room always equals that of the external air. In the usual notation, the (inner, thermal) energy is, per unit mass,

$$E = C_V T.$$

(An additive constant may be neglected.) Then the energy content is, per unit of volume,

$$E_1 = C_V \rho T,$$

[ρ = density] or, taking into account the equation of state,

$$\frac{p}{\rho} = RT,$$

we have

$$E_1 = \frac{C_V p}{R}.$$

"For air at atmospheric pressure,

$$E_1 = 0.0604 \text{ cal}\cdot \text{cm}^{-3} = 60.4 \text{ cal}\cdot \text{m}^{-3}$$

[24]What follows is reproduced verbatim from the original article (written in English) in *Supplement to NATURE*, May 21, 1938, pp. 908–909, except that the symbolism is changed to conform with that used in this book. *Trans.*

[= $2.528 \cdot 10^5$ joules/m^5]. *The energy content of the room is thus independent of the temperature*, solely determined by the state of the barometer. The whole of the energy imported by the heating escapes through the pores of the walls of the room to the outside air.

"I fetch a bottle of claret from the cold cellar and put it to be tempered in the warm room. It becomes warmer, but the increased energy content is not borrowed from the air of the room but is brought in from outside.

"Then why do we have heating? For the same reason that life on earth needs the radiation of the sun. But this does not exist on the incident energy, for the latter apart from a negligible amount is re-radiated, just as a man, in spite of continual absorption of nourishment, maintains a constant body-weight. Our conditions of existence require a determinate degree of temperature, and for the maintenance of this there is needed not addition of energy but addition of entropy.

"As a student I read with advantage a small book by F. Wald[25] entitled 'The Mistress of the World and her Shadow'. These meant energy and entropy. In the course of our advancing knowledge the two seem to me to have changed places. In the huge manufactory of natural processes, the principle of entropy occupies the position of manager, for it dictates the manner and method of the whole business, whilst the principle of energy merely does the book-keeping, balancing credits and debits.

R. Emden. Kempterstrasse 5, Zürich."

Sommerfeld refined Emden's argument, concluding that the energy density of a room does not in fact remain constant, but actually *decreases* with heating—which goes to show all the more the validity of the conclusion that entropy plays the leading role over that of energy.

"The mistress of the world and her shadow"

The book by Wald that Emden refers to is not the only book with this title. As a child I read a book with the same title by Berthold Auerbach.[26] Thanks to popularizers such metaphors have retained their currency. But Emden's reversal of the metaphor is correct.

Energy is called "mistress of the world" because everything that happens in the world does so via changes of one form of energy into another. Einstein showed that a quantity of matter of mass m is equivalent to an amount of energy given by

$$E = mc^2,$$

where c is the speed of light. Energy is in an essential way contained in the material of the world—matter and fields. Most of this energy is liberated and used only in certain of the transformation processes of atomic nuclei—whence atomic

[25] František Wald (1861–1930), Czech chemist. *Trans.*

[26] Berthold Auerbach (1812–1882), German-Jewish poet and author. *Trans.*

energy. The other forms of energy—potential and kinetic, thermal and chemical, electrical and magnetic—are the direct sources of the work carried out in nature and technology. Work is done when one of these forms of energy is transformed into another.

Entropy is called "the shadow" of the mistress of the world because it can be used as a measure of the depreciation of energy, if we understand the value of energy to lie in its availability for transformation into useful work. As we saw earlier, the maximum amount of useful work is equal to the decrease in Helmholtz free energy (see (3.4)). But the Helmholtz free energy is the change in internal energy less the heat absorbed (see (3.5)):

$$F = E - TS.$$

Thus the greater the heat absorbed, or, equivalently, the greater the increase in entropy, the less energy available for doing useful work, that is, the less valuable the internal energy E. As mentioned earlier, in purely mechanical processes all of the energy is available for doing work, but in processes involving heat exchange—such as mechanical processes where friction occurs—some of the energy is transformed into heat, that is, into entropy times the temperature.

Why was Emden right?

Answer: Because the direction of flow of all real processes is determined by the direction of change in entropy. As we said earlier, all real processes are irreversible, so that in an isolated system they will proceed in the direction of increasing entropy. However, this does not mean that entropy cannot decrease. The law of increasing entropy holds only for isolated systems. In open systems, that is, in systems in which matter and energy are exchanged with the surrounding medium, the situation is very different. All phenomena of the biosphere, that is, occurring in living nature, involve changes in entropy. Under normal conditions, the mass and supply of energy of, for instance, a human organism remain constant; they are constantly maintained through breathing and eating. However, this replenishment involves a decrease in entropy rather than an increase in energy. An important consequence of the second law is the position set out in Emden's article, also formulated in Erwin Schrödinger's famous book[27] *What is life (from the point of view of physics)?*: *A living organism feeds on negative entropy.* In Chapter 6 we explain what this means exactly.

Thus Emden thinks that we should interchange the places of entropy and energy, shadow and mistress—just as in Hans Christian Andersen's fairy tale "The shadow", turned into a marvelous play by Evgeniĭ Schwartz.[28]

In fact the metaphor of mistress and shadow is of no great significance. It would perhaps be better to abandon it and refer to energy and entropy neither as

[27]Published as *What is life?* in 1944. *Trans.*

[28]Evgeniĭ Schwartz (1896–1958), Russian playwright. *Trans.*

mistress nor shadow. What we do retain is the discovery that entropy is just as important as energy, and especially so in cosmology and biology—which is why time flows the way it does, from the past into the future.

Chapter 4

Entropy and probability

If you aspire to conceive a potent idea
In the name of the advancement of knowledge,
You must first carry out
A statistical summation.

This will help you in everything,
It will reveal a lighted path in the gloom.
But, try as you might, without it
The essence of phenomena will remain inaccessible.

Boltzmann's formula

Thus far we have studied only phenomenological physics, that is, thermodynamics, where our systems are described in terms of state functions such as energy, enthalpy, entropy, and free energy. In particular, we discovered that entropy increases in spontaneously evolving processes. But why should this be so?

The answer to this question is contained in Boltzmann's formula

$$S = k \ln P, \tag{4.1}$$

where P denotes the so-called "statistical weight"[1] of the current state of the system in question, and k is a constant called *Boltzmann's constant.* It is the ratio of the gas constant $R = 8.31$ joules/(mole·°K) to the Avogadro number (the number of molecules in a mole of gas) $N_A = 6.06 \cdot 10^{23}$ per mole:

$$k = \frac{R}{N_A} = 1.38 \cdot 10^{-23} \text{ joules/°K}. \tag{4.2}$$

Although it is not possible to give a fully rigorous derivation of this celebrated formula in such a popular account as this, we shall nonetheless attempt to show

[1] Defined below. *Trans.*

how the entropy of a gas in a given state and the probability of that state must be related by such a formula.

First we must understand what is meant by P, the "statistical weight" of the state of the system. Clearly, this cannot be the usual probability of one outcome out of several possible outcomes, such as when a die is cast and the probability of the outcome 3 is $q = 1/6$.[2] Note that the probability is less than 1, as it must be since the sum of the probabilities of each of the outcomes 1,2,3,4,5,6 must be 1.[3] For similar reasons, probabilities are generally less than 1, and since the logarithm of a (positive) number less than 1 is negative, the "statistical weight" P appearing in formula (4.1) cannot be ordinary probability.

The *statistical weight* of a state of a system is defined to be the number of ways that the state can be realized. Since the state of a thrown cubical die is the same regardless of the numerals on its faces, the statistical weight of a cast die is 6, $P = 6$.

If we throw two dice, then the number of outcomes is $6 \cdot 6 = 36$. The probability of, for instance, obtaining a 3 on one die and a 4 on the other[4] is equal to the product of the two separate probabilities, since these two outcomes are independent.[5] It follows that here the number P, again in view of the independence of the two events, is equal to the product of the numbers P_1 and P_2:

$$P = P_1 \cdot P_2. \tag{4.3}$$

Boltzmann's formula is then made plausible by the fact that it yields the additivity of entropies directly:

$$S = k \ln P = k \ln(P_1 P_2) = k \ln P_1 + k \ln P_2 = S_1 + S_2. \tag{4.4}$$

Suppose now that we have four molecules distributed among two boxes, as in Figure 4.1. How many different states can there be? If we assume the molecules identical, then clearly the number of states is five: $4|0, 3|1, 2|2, 1|3, 0|4$. The number of ways in which these states can be realized, that is, their statistical weights, are different. If we assume the molecules distinguishable by being differently numbered or colored, then it can be seen from Figure 4.1 that the statistical weights are respectively $1, 4, 6, 4, 1$. The most likely distribution is the uniform one $2|2$, with two molecules in each box.

---

[2]Given an experiment with a finite number N of mutually exclusive and equally likely outcomes, the *probability* of an event A, that is, a subset consisting of n outcomes, is defined to be n/N. If the probability space of all possible outcomes is infinite, then an appropriate "probability measure" has to be defined on it so that the measure of the whole space is 1, and then the probability of an event, that is, a measurable subset, is just its measure. *Trans.*

[3]That is, the probability that *some* outcome occurs must be 1. *Trans.*

[4]Here the dice are assumed distinguished from one another. *Trans.*

[5]Two events A and B are said to be *independent* if $\text{prob}(A \cap B) = \text{prob}A \cdot \text{prob}B$, that is, if the probability of both events occurring is equal to the product of their separate probabilities. This captures in precise form the idea that the occurrence of either event should not change the probability of the other. *Trans.*

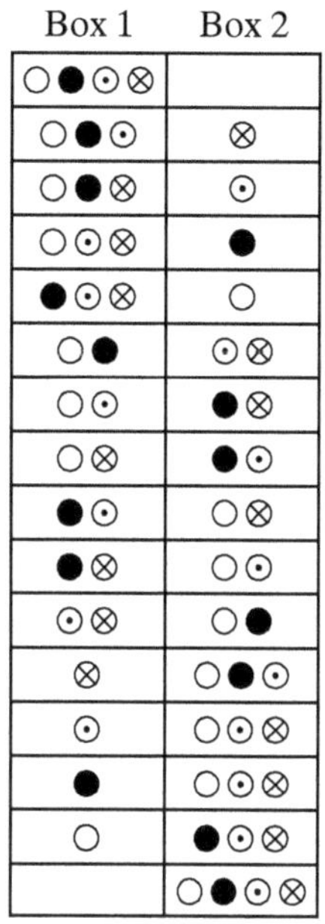

Figure 4.1: The possible states of four particles distributed over two boxes.

Where do the numbers $1, 4, 6, 4, 1$ come from? They are the binomial coefficients

$$\frac{4!}{4!0!} = 1, \ \frac{4!}{3!1!} = 4, \ \frac{4!}{2!2!} = 6, \ \frac{4!}{1!3!} = 4, \ \frac{4!}{0!4!} = 1.$$

We remind the reader that for any positive integer N, the symbol $N!$, read "N-factorial", denotes the product of all numbers from 1 to N:

$$N! = 1 \cdot 2 \cdot 3 \cdots N,$$

and

$$0! = 1.$$

Thus $1! = 1, 2! = 2, 3! = 6, 4! = 24$, and so on.

In the general case of N molecules distributed over two boxes, the number of different ways of obtaining the distribution $N_1|N_2$, where $N_1 + N_2 = N$, is

$$P = \frac{N!}{N_1!N_2!} = \frac{N!}{N_1!(N - N_1)!}. \tag{4.5}$$

This is not difficult to see: the numerator $N!$ is the total number of permutations (ordered linear arrangements) of the N molecules. If we imagine the first N_1 molecules in each arrangement as being in the first box, and the rest in the second, then since the order within each box is immaterial, and there are $N_1!N_2!$

permutations fixing each partition of the N molecules into two subsets of sizes N_1 and N_2 respectively, we need to divide by $N_1!N_2!$.

We now use Boltzmann's formula to compute the change in entropy due to the mixing of two gases consisting of N_1 and N_2 molecules respectively.

Before being mixed, the gases are assumed to be separated by a partition. We distinguish their states only by location: the N_1 molecules of the first gas are all in the left half of the container, and the N_2 molecules of the second in the right half. The respective statistical weights are then

$$P_1 = N_1!,\ P_2 = N_2!,$$

and, since entropies are additive for the unmixed system, its entropy is given by

$$S = S_1 + S_2 = k(\ln N_1! + \ln N_2!).$$

When the partition is removed, the gases mix. The entropy of the mixture is

$$S' = k\ln(N_1 + N_2)! = k\ln N!.$$

The mixing entropy is therefore

$$\Delta S = S' - S = k\ln\frac{N!}{N_1!(N-N_1)!}. \tag{4.6}$$

Thus the mixing entropy is given by k times the logarithm of the expression (4.5), so in order to estimate it we need a good approximation to the factorials of large numbers, where by "large" we mean of an order much larger than 1.

Stirling's formula

From $N! = 1\cdot 2\cdot 3\cdots N$, we infer that

$$\ln N! = \ln 1 + \ln 2 + \ln 3 + \cdots + \ln N = \sum_{i=1}^{N}\ln i. \tag{4.7}$$

The function $\ln N$ increases more and more slowly with increasing N, since the difference $\ln(N+1) - \ln N = \ln(1+1/N)$ decreases with increasing N. Hence for large N the area under the graph of $y = \ln x$ affords a good approximation of $\ln N!$:

$$\ln N! \approx \int_1^N \ln x dx. \tag{4.8}$$

This integral can be evaluated using "integration by parts". We recall how this is done. For functions u and v of x, we have

$$d(uv) = udv + vdu.$$

Hence

$$\int_a^b u dv = \int_a^b d(uv) - \int_a^b v du = uv \mid_a^b - \int_a^b v du.$$

In our case, we take $u = \ln x$ and $v = x$, obtaining

$$\begin{aligned}\int_1^N \ln x dx = x \ln x \mid_1^N &- \int_1^N x d(\ln x) \\ &= N \ln N - \int_1^N x \frac{dx}{x} = N \ln N - (N - 1).\end{aligned}$$

Since N and $N \ln N$ are large compared with 1, we can neglect the term -1. Thus we end up with the approximation

$$\ln N! \approx N \ln N - N = N \ln \frac{N}{e}, \tag{4.9}$$

yielding in turn the rough approximation

$$N! \approx \left(\frac{N}{e}\right)^N. \tag{4.10}$$

A more accurate approximation is given by Stirling's formula

$$N! \approx (2\pi N)^{1/2} \left(\frac{N}{e}\right)^N. \tag{4.11}$$

Taking logarithms, we obtain

$$\ln N! \approx N \ln N - N + \frac{1}{2} \ln N + \frac{1}{2} \ln 2\pi, \tag{4.12}$$

which is not so much better than (4.9), since for large N in (4.12) we can neglect the terms $\frac{1}{2} \ln N$ and $\frac{1}{2} \ln 2\pi$ in comparison with $N \ln N$ and N. Thus we may use (4.9) as an approximation of $\ln N!$.

Having settled on an approximation of the logarithm of factorials, we return to the formula (4.6). That formula yields, via the approximation (4.9),

$$\begin{aligned}\Delta S &\approx k(N \ln N - N - N_1 \ln N_1 - N_2 \ln N_2 + N_1 + N_2) \\ &= k[(N_1 + N_2) \ln(N_1 + N_2) - N_1 \ln N_1 - N_2 \ln N_2],\end{aligned}$$

or

$$\Delta S \approx k \left(N_1 \ln \frac{N_1 + N_2}{N_1} + N_2 \ln \frac{N_1 + N_2}{N_2} \right). \tag{4.13}$$

If we express the numbers of molecules in terms of the numbers n_1, n_2 of moles:

$$N_1 = N_A n_1, \; N_2 = N_A n_2,$$

where N_A is Avogadro's number, (4.13) becomes

$$\Delta S = kN_A \left(n_1 \ln \frac{n_1 + n_2}{n_1} + n_2 \ln \frac{n_1 + n_2}{n_2} \right). \tag{4.14}$$

Since $kN_A = R$, we have arrived via Boltzmann's formula at the former expression (2.24) for the mixing entropy, derived by means of a completely different argument.

The meaning of Boltzmann's formula

Thus we have provided considerable evidence for, though of course not proved, Boltzmann's formula, expressing entropy as a constant times the logarithm of the statistical weight of the system. A great many things follow from this.

Under equilibrium conditions, entropy is not in any way remarkable. It is a function of the state of the system which can be measured experimentally (as well as computed theoretically using Boltzmann's formula, as we shall see below). However, as soon as an isolated system deviates from equilibrium, a remarkable property of entropy emerges, namely, its propensity to always increase to a maximum.

This property can be demonstrated using the relation between S and the statistical weight P. We may rewrite Boltzmann's formula in the exponential form

$$P = e^{S/k}, \tag{4.15}$$

which shows that the probability of a particular state increases exponentially with its entropy. (Note that the entropy S can be determined experimentally, while P can be calculated from its definition.) Thus the increase in entropy in an irreversible process entails a change to a more probable state. This is confirmed by the fact that a disordered state is more likely than an ordered one.

Suppose we have initially, as earlier, two bodies at different temperatures. This implies a certain order. If the temperatures of the bodies are equalized by means of heat conduction—the flow of heat from the hotter to the colder body—, then this order is destroyed. The same sort of thing occurs when gases or liquids are mixed. Again, the unrestricted expansion of a gas such as hydrogen or carbon dioxide, when released from a balloon containing it, represents increasing disorder. While the gas was trapped in the balloon, it occupied little space. On being released it expands freely, for the simple reason that the probability is greater that it should occupy a larger volume than a smaller, that is, that the statistical weight of the state of the gas occupying a larger volume is greater than that of its state when occupying a smaller volume. For a given system, a less ordered state has larger statistical weight since it can be realized in more ways than a more ordered state.

If no conscious effort is made to arrange the books and papers on a desk neatly, they will end up in a state of disorder, as a result their being randomly moved about—randomly, since unsystematically.

In these examples order is created artificially, while disorder arises spontaneously, being associated with greater probability, greater entropy. Thus one might say that entropy is a measure of the amount of disorder of a state of a system.

The aim of people's and animal's rational activity is the overcoming of disorder. For example, members of a productive farm have to struggle to thwart spontaneous processes such as soil erosion, deterioration of crops, and so on. Here too is entropy more significant than energy.

From all this we see that the second law of thermodynamics is of a quite different sort from the first. The first law holds in all cases: energy is conserved in interactions of elementary particles just as it is in macroscopic systems.

It is worth mentioning that in the 1930s the idea arose that energy might be conserved only on average, and not necessarily in certain individual processes involving elementary particles. As might be expected, this idea owed much to the nature of the second law. The American physicist Shankland[6], in his investigations of the interactions of photons and electrons, thought that he had found evidence for the non-conservation of energy in such basic processes. Even the great Niels Bohr then allowed the possibility that the law of conservation of energy might not apply to individual events in the microcosm, but hold only on average. However Shankland's experiments were soon shown to be flawed, and since then the rigorously deterministic nature of the first law has remained unquestioned.

Incidentally, this law, together with the law of conservation of momentum, served as the basis of a great discovery, that of the neutrino. The theory of the α-decay[7] of radium was created by G. A. Gamow[8] in 1928. Here no contravention of the law of conservation of energy arose. However, β-decay, involving the radioactive emission of an electron or positron from an atomic nucleus, represented at that time a seemingly insoluble puzzle: the conservation laws appeared not to hold! Gamow, who was somewhat of a jokester, made a rubber stamp of a skull over two crossed βs (Figure 4.2), which he used to stamp on his offprints. This symbolized the difficulties with the theory of β-decay.

This Gordian knot was cut in 1933 by Wolfgang Pauli, who introduced a new particle, the neutrino, in order to save the conservation laws. The neutrino was later detected experimentally.

Unlike the first law, the law asserting the impossibility of a perpetual motion machine of the second kind—the law of increase of entropy—is not deterministic, but statistical, probabilistic, since the impossibility of a perpetual motion machine of the second kind is a consequence of its improbability. Let us estimate, for instance, the probability that all molecules in a container of volume V spontaneously gather together in one half of the container. The probability of a single molecule's being found in, say, the right half of the container is $1/2$. If there are N

[6]Robert Sherwood Shankland (1908–1982), American physicist and historian of science. *Trans.*

[7]An α-particle is identical to a helium nucleus, consisting of two protons and two neutrons bound together. *Trans.*

[8]George Gamow (1904–1968), Russian physicist, in the U.S. from 1934. *Trans.*

Figure 4.2: Symbol of the theoretical difficulties associated with β-decay.

molecules in all, then the probability that all N will gather together in that half of the container is the product of the probabilities for the individual molecules, since these are independent events. Suppose the volume V is a cubic centimeter. At ordinary temperature and pressure, a cubic centimeter of a gas contains approximately $2.7 \cdot 10^{19}$ molecules (Loschmidt's number). Hence the desired probability is about

$$\left(\frac{1}{2}\right)^{2.7\cdot 10^{19}},$$

which is vanishingly small. Violation of the second law would require a highly improbable event such as this to take place—compression of a gas without doing any work on it.

The discoveries of Boltzmann and Gibbs, the creators of statistical physics, heralded a scientific revolution, a breakthrough into an entirely new field. Of course, as is always the case in science, this revolution did not arise out of nowhere. The ground was prepared by Gibbs' and Boltzmann's predecessors: Carnot, Clausius, Thomson, and Maxwell.

The main innovation consisted in the probabilistic, rather than deterministic, nature of the new statistical laws. We said earlier that it is impossible for the water in a kettle submerged in a bucket of water to come to a boil at the same time as the water in the bucket freeze. The sense of the word "impossible" here, is that of vanishingly small probability, not that the event in question is categorically ruled out—passing strange though it would be if it occurred. After an extremely large number of trials, it may come to pass.[9] It is just that a spontaneous drop, rather than rise, in entropy is a rare—extremely rare!—occurrence. We shall look further into this issue in the sequel. For now, suffice it to say that, as the above calculation shows, in a macroscopic system exceptions to the probabilistic second law are very few and far between. Nevertheless, such exceptions occur, and we witness them every day—for example, in the statistics behind the blueness of the sky. We shall explain this in the next chapter.

[9]In fact, it follows from Poincaré's recurrence theorem and a theorem of Liouville, that it is almost certain that the unlikely event in question will *eventually* occur. See, for example, the popular book *Easy as π?* by O.A. Ivanov, Springer, New York, etc., 1998, pp. 105–110. *Trans.*

In the classical account of statistical mechanics written by Joseph Mayer and his wife Maria Goeppert-Mayer[10] a discussion of these issues is preceded by the following epigraph:

"Never?" "No, never."
"Absolutely never?" "Well, hardly ever."

The fusion of a crystal and the evaporation of a liquid

We discussed the fusion of crystals and evaporation of liquids in the preceding chapter. We are now in a position to understand the physical meaning of such phase transitions.

We saw in Chapter 3 that phase transitions involve changes ΔH in enthalpy and ΔS in entropy. Such transitions are said to be "of the first kind." When a crystal melts, its disordering as it assumes liquid form might be called "entropically advantageous." The disordered state of the constituent atoms or molecules is more probable than their ordered state, since it can be realized by a greater number of arrangements of these particles than the ordered crystalline state. Hence the entropy of the liquid state of the substance is greater than its entropy in the crystalline state. On the other hand, the enthalpy of the liquid is less than the enthalpy of the crystal: in order to fuse the crystal one must break some of the interatomic or intermolecular bonds. Thus the crystalline state is advantageous energy- and enthalpy-wise, but disadvantageous entropy-wise. Fusion begins precisely when the entropic contribution to the difference in free energies of crystal and liquid exactly compensates the enthalpic contribution:

$$T_{melting}\Delta S = \Delta H.$$

At higher temperatures $T > T_{melting}$, we have

$$T\Delta S > \Delta H,$$

so that the entropic contribution exceeds ΔH, and the more probable liquid state is the more stable. At lower temperatures $T < T_{melting}$,

$$T\Delta S < \Delta H,$$

and the enthalpic contribution dominates—the crystal state is the more stable.

Similar considerations hold for evaporation. The gaseous state is more likely than the liquid state, but the enthalpy of the liquid is less because of the interactions between its particles.

---

[10] Maria Goeppert-Mayer (1906–1972), German-born, American theoretical physicist, Nobel laureate in 1963. *Trans.*

But why is ΔH much larger at evaporation than at fusion? In Chapter 3 we quoted the values for water:

$$\Delta H_{melting} = 5982 \text{ joules/mole}, \ \Delta H_{evaporation} = 40613 \text{ joules/mole}.$$

For carbon tetrachloride they are

$$\Delta H_{melting} = 2416.2 \text{ joules/mole, and } \Delta H_{evaporation} = 32410.9 \text{ joules/mole}.$$

The explanation lies in the fact that in a liquid not all of the bonds between the particles in the crystalline state have been broken. Like a crystal, a liquid is a condensed body; the densities of liquid and solid are not markedly different. However on evaporation, all bonds between the particles are broken, and this naturally requires a considerable input of energy. And the change in entropy is also greater.

Entropic forces

In Chapter 3, we saw that the reason for gasoline's not dissolving in water is that the hydrocarbon molecules are repelled from the water by entropic forces, and that an albumin chain bundles itself into a globule for similar reasons. We also saw that the elasticities of an ideal gas and a rubber band are determined by changes in entropy rather than internal energy. Now that we know the statistical nature of entropy, we are in a position to understand just why these entropic forces arise.

On mixing a hydrocarbon with water, the entropy goes down. This must mean that the state of the mixture, which we indicate schematically by (using the hydrocarbon benzol, C_6H_6)

$$\begin{array}{ccc} H_2O & \cdots & C_6H_6 \\ C_6H_6 & \cdots & H_2O, \end{array}$$

has lower probability than the state of unmixed water and hydrocarbon:

$$\begin{array}{ccc} C_6H_6 & \cdots & C_6H_6 \\ H_2O & \cdots & H_2O. \end{array}$$

Yet it might seem that the probability of the first of these two states should be *more* probable than the second, since it is a mixture, and we have seen that mixing entropy is positive. What's going on here?

The point is that we cannot consider the dissolving of a hydrocarbon in water as a simple mixing like, say, the mixing of red and blue glass beads. Water is a complex liquid, as is shown by its well-known properties, one of which is that in its solid form, as ice, it is lighter than in its liquid form; the maximum density of water occurs not at 0°C, the melting point of ice, but at 4°C (at atmospheric pressure). The reason for this is that water is an "associative" liquid, meaning that between its molecules there are so-called "hydrogen bonds", as shown by the

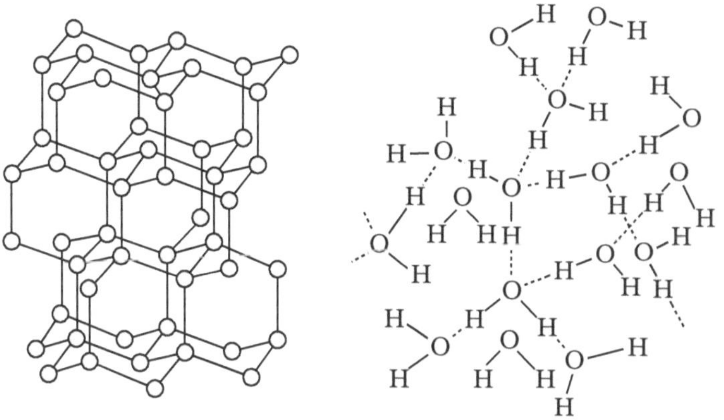

(a) (b)

Figure 4.3: The structure of ice (a) and of liquid water (b). In (a) the small circles denote the molecules of H_2O, and the straight line segments the hydrogen bonds between them. In (b) the solid lines denote chemical bonds O—H and the dashed lines hydrogen bonds.

dashed lines in Figure 4.3 (b). These bonds are much weaker than chemical bonds. The energy of the chemical bond O—H, that is, the amount of work needed to break it, is about 463 kilojoules/mole, whereas the energy of the hydrogen bond $OH\cdots O$ in water is of the order of 12 kilojoules/mole.

An ice crystal is light because of its spacious structure; each of its molecules neighbours on just four other molecules, arranged in the form of a tetrahedron (Figure 4.3(a)). Thus a large proportion of the space occupied by the crystalline lattice is empty. In liquid water the icelike lattice has been slackened; some of the hydrogen bonds connecting each molecule of the ice to its four neighbours are broken, so that in liquid water there are molecules with four, three, two, one, or even no bonds with adjacent molecules (Figure 4.3(b)). This is why water is less ordered yet denser than ice—the loosening of the rigid crystalline lattice allows water molecules into its empty spaces.

The decrease in entropy that occurs if a hydrocarbon is dissolved in water indicates that greater order must have come about. What happens is that water molecules surrounding a molecule of the hydrocarbon tend to behave as they do in crystalline ice, forming, so to speak, a microscopic iceberg around each hydrocarbon molecule. However, such an increase in order is an improbable event, while its destruction is probable. Thus the separation of the mixture into two layers is more likely than the hydrocarbon's dissolving in the water. As we saw earlier, this "force" of separation explains the properties of soap and the globular structure of naturally occurring albumin.

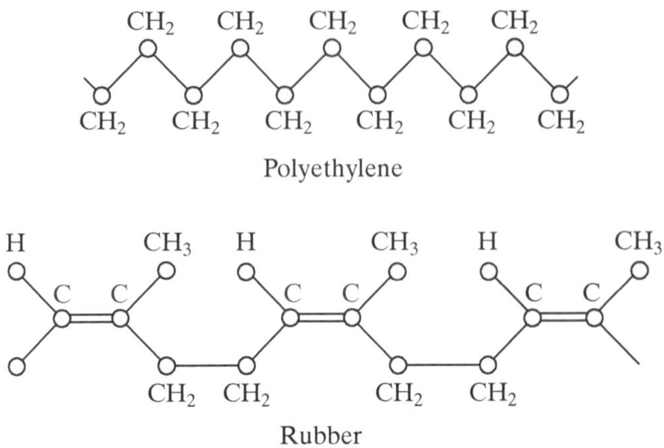

Figure 4.4: Molecular structures of polethylene and rubber.

Whence comes the elastic force of an ideal gas resisting compression? It is clear that a compressed state of a gas is less likely than a more expanded one, so the smaller the volume occupied by the gas the greater the orderliness, whence the tendency to expand to a state of greater entropy. And in terms purely of molecular physics, the entropic elastic force of the gas is intuitively comprehensible: when a gas is compressed, the number of impacts of its molecules against the piston increases, so that the piston encounters ever increasing resistance.

All this is clear, but what happens in the case of rubber? Precisely why is it that when we stretch a rubber band or tube there comes into being an entropic elastic force tending to restore the original state? What does a rubber band have in common with an ideal gas from the point of view of molecular physics?

The entropic force of an ideal gas is determined by statistical, probabilistic properties. The gas consists of a swarm of independently moving, mutually non-interacting molecules. So obviously rubber must also contain kinetically independent parts subject only to the laws of probability and statistics. In order to locate these parts we need to examine the molecular structure of rubber; this is shown in Figure 4.4. Both natural and synthetic rubber are made of long polymer chains. Since the properties of rubber—above all its great elasticity—are, in principle, shared by a great many polymers, we shall simplify our task by considering instead one of the simplest—if not the simplest—of all polymers, namely polyethylene, whose molecular structure is also shown in Figure 4.4. Ordinary polyethylene, familiar to everybody since it is used for making plastic bags and film, does not in fact have the same physical properties as rubber. This is so because it is partly crystallized. However if conditions are right, polyethylene and rubber do behave similarly.

Now a polymer chain is formed by means of single chemical bonds of the form —CH_2—CH_2—. Such bonds admit rotations that do not require especially large

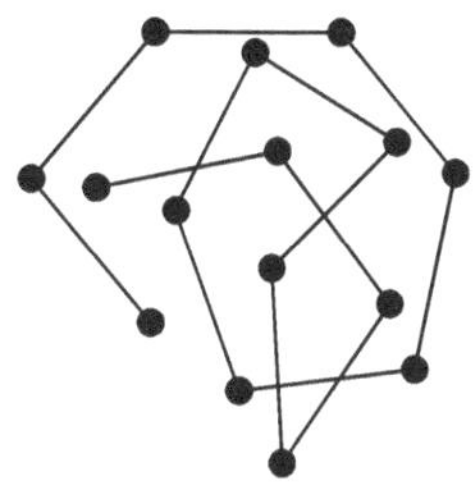

Figure 4.5: Model of a polymer ball.

amounts of energy. Rotation about the C—C bond in a molecule of ethane H_3C—CH_3 requires about 12 kilojoules/mole. The thermal energy per degree of freedom in a molecule at room temperature that goes into a rotation—in particular one about a C—C bond—is about $N_A kT = RT = 2.5$ kilojoules/mole, while the total amount of energy needed to achieve such a rotation is about five times greater. Nevertheless, at room temperature rotations do occur all the time—after all kT represents the average energy, and there will be many molecules with much higher energy. Therefore, at room temperature, macromolecules, that is, polymer chains in solution, undergo all possible internal rotations about their single bonds—in fact this motion is what constitutes thermal energy. As a result, a macromolecule in solution rolls up into a ball. One must not confuse such a polymer ball with an albumin globule. A polymer ball is a loose formation permeable by the solvent, whereas an albumin globule is a closely packed, dense body. Figure 4.5 depicts schematically the structure of a polymer ball as obtained from a loosely coupled hinged chain. Such a depiction is appropriate since if rotations can occur about certain single bonds, then the chain may be represented by a sequence of loosely coupled segments. The state of being rolled up in a ball is more likely since it can be realized in many more ways than the state of a chain that is stretched out. The loosely coupled segments may be considered as the independent kinetic elements of the rubber. When a rubber band is stretched, its state changes from a more likely one, where its chains are rolled up into balls, to a less likely one where they are stretched out. Entropy is thereby decreased and an elastic force proportional to the temperature arises.

Thus the similarity in elastic behaviour between an ideal gas and a rubber band comes down to the fact that in both cases the system consists of a large number of independent kinetic elements, namely, molecules and chain segments respectively. Such a system is subject to statistics, its behaviour is of a probabilistic character.

Entropy of equilibrium states

An isolated system in a state of equilibrium has maximum entropy. How can this entropy be expressed statistically?

From Boltzmann's formula we see that we should look for the maximum value of the statistical weight P. But in the general case how does one go about finding P? Suppose we have a physical system consisting of atoms or molecules, each of which can exist in any state of a certain possible set. The general formula for the statistical weight is

$$P = \frac{N!}{N_1!N_2!\cdots N_M!}, \tag{4.16}$$

where $N_1, N_2, \ldots, N_M$ denote the numbers of particles—atoms or molecules—in states $1, 2, \ldots, M$ respectively, and N is the total number of particles. (This represents a generalization of the formula (4.5).) Thus

$$N = \sum_{i=1}^{M} N_i = N_1 + N_2 + \cdots + N_M, \tag{4.17}$$

and the total energy of the system in this state is

$$E = \sum_{i=1}^{M} N_i E_i, \tag{4.18}$$

where E_i is the energy of a particle in the Mth state. Thus our problem here is to find the maximum value of the expression

$$S = k \ln P = k\left(\ln(N!) - \sum_{i=1}^{M} \ln(N_i!)\right).$$

From the approximations $\ln(N!) \approx N\ln N - N$, $\ln(N_i!) \approx N_i \ln N_i - N_i$ (see (4.9)), it follows that

$$S = k \ln P \approx k\left(N\ln N - \sum_{i=1}^{M} N_i \ln N_i\right). \tag{4.19}$$

Thus we wish to maximize the right-hand expression in (4.19) subject to the conditions (4.17) and (4.18). Higher mathematics enables one to solve this problem easily; however we shall skip the mathematics and go right to the answer. It turns out that the most likely energy distribution among the particles, that is, the distribution corresponding to an equilibrium state, or, equivalently, to maximal entropy, is given by the formula

$$p_i = \frac{N_i}{N} = \frac{e^{-E_i/kT}}{\sum_{i=1}^{M} e^{-E_i/kT}}, \tag{4.20}$$

where p_i is the probability that a randomly chosen molecule is in the state with energy E_i. It is immediate that

$$\sum_{i=1}^{M} p_i = 1, \tag{4.21}$$

since the probability that a molecule is in one of the possible M states must be unity. The probability distribution (4.20) is called *Boltzmann's distribution.* It is valid for systems consisting of atoms, molecules, or other particles that can be mentally enumerated, or distinguished from one another in some other way; this is the basic assumption underlying the formula (4.21).[11] The formula (4.16) for P (see also (4.5)) is valid only under this assumption.

In the Boltzmann distribution there appears the important physical quantity

$$Z = \sum_{i=1}^{M} e^{-E_i/kT}, \tag{4.22}$$

called the *canonical partition function* or *statistical sum* of the system. Thus we may rewrite (4.20) as

$$p_i = \frac{e^{-E_i/kT}}{Z}. \tag{4.23}$$

All of the chief thermodynamical characteristics of a body—its energy, entropy, and free energy—can be expressed in terms of the quantity Z. Thus it is not difficult to show that for one mole of a gas we have

$$E = RT^2 \frac{d \ln Z}{dT}, \tag{4.24}$$

and

$$S = \frac{E}{T} + R \ln Z. \tag{4.25}$$

From (4.24) and (4.25) the expression for the Gibbs free energy in terms of Z follows:

$$F = E - TS = -RT \ln Z. \tag{4.26}$$

Thus of the three quantities, the free energy has the simplest expression in terms of the partition function. Exponentiation of both sides of (4.26) yields

$$Z = e^{-F/RT}, \tag{4.27}$$

[11] For a general closed system consisting of N particles the statistical weight of a state cannot be defined by simple counting, since there are infinitely many positions that the particles may take up, and infinitely many velocities. Instead one works in $6N$-dimensional *phase space*, each of whose points is a $6N$-tuple made up of $3N$ position coordinates and $3N$ velocity coordinates. The phase space is divided up into regions such that distinctions within each region are "macroscopically indistinguishable". The statistical weight of a state is then defined by a suitable volume-like "measure" that assigns the same weight to each of the "microscopic" regions configured in agreement with the values of the parameters determining the given macroscopic state. *Trans.*

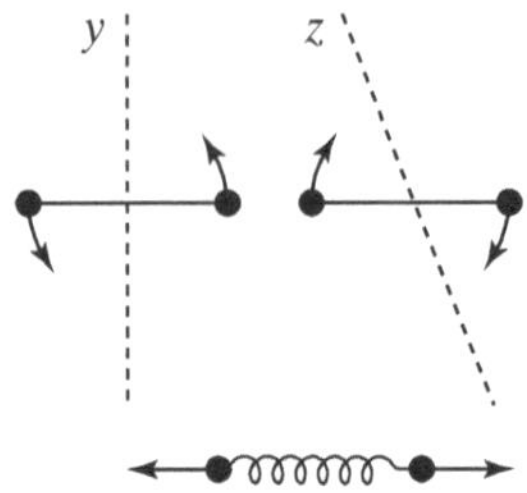

Figure 4.6: Rotations and oscillation of a two-atom molecule.

an exponential expression for the partition function in terms of the free energy.

These formulae apply to a gas; however it is possible to find partition functions also for solids and liquids. While this is not always so easy, the effort is worthwhile, since knowledge of the partition function allows one to easily infer the basic thermodynamical quantities. However, to obtain the partition function one must know all possible values E_i of the energy of all conceivable states of the components of the system.

A little quantum mechanics

Let us suppose we are carrying out calculations pertaining to a gas consisting of two-atom molecules, such as nitrogen, N_2. What are the energy levels characteristic of these molecules?

Nitrogen molecules can move rectilinearly in any direction, so that they possess three degrees of freedom as far as their rectilinear motion is concerned. A nitrogen molecule can also rotate about any axis perpendicular to the bond line of the two atoms comprising it. (Rotation about the axis coincident with this line changes nothing and does not involve outlays of energy.) Hence each nitrogen molecule possesses two rotational degrees of freedom. Finally, the two atoms of a molecule can oscillate back and forth in the direction of the bond line—as if the chemical bond were a spring. Hence there is also one vibrational degree of freedom. These rotational and oscillatory motions are represented schematically in Figure 4.6.

In addition to all that, the electrons of a molecule have a characteristic set of energy levels, just as they do in an atom.

Thus there are in all four distinct contributions to the total energy of a nitrogen molecule—from its rectilinear motion, rotation, oscillation, and electrons, respectively:

$$E = E_{rect} + E_{rot} + E_{vib} + E_{el}. \tag{4.28}$$

Using quantum mechanics, the physics of the microcosm, one can calculate these contributions with great precision.

Observe first, however, that since the energy levels appear in the partition function as exponents, we have

$$\begin{aligned} Z &= \sum_i e^{-E_i/kT} \\ &= \sum_a e^{-E_{rect_a}/kT} \sum_b e^{-E_{rot_b}/kT} \sum_c e^{-E_{vib_c}/kT} \sum_d e^{-E_{el_d}/kT}, \end{aligned}$$

or

$$Z = Z_{rect} Z_{rot} Z_{vib} Z_{el}. \tag{4.29}$$

Now under normal conditions, Z_{el} is practically equal to unity. The reason behind this is quantum mechanical: the energy levels of electrons in atoms and molecules are discrete, in the sense that E_{el_d} can take on only certain definite values

$$E_{el_1}, E_{el_2}, \ldots.$$

These energies can be computed starting from the fundamental state of the molecule, in which the electron is not excited, that is, we may put $E_{el_1} = 0$. On the other hand, the energies of the excited states of electrons are much greater than kT, so that the corresponding terms in the sum Z_{el} are close to zero. We conclude that $Z_{el} \approx 1$.

The energy levels of the rotational and oscillatory motions of the molecule are also discrete, but are of the same order of magnitude as kT. The energy of the rectilinear motion of the molecule is not discrete and depends on the temperature.

Taking logarithms, we obtain

$$\ln Z = \ln Z_{rect} + \ln Z_{rot} + \ln Z_{vib}. \tag{4.30}$$

Since the energies associated with the various kinds of motion add (see (4.28)), we obtain from (4.25) and (4.30) that the entropies also add:

$$S = S_{rect} + S_{rot} + S_{vib}. \tag{4.31}$$

From the values of the relevant energies obtained using quantum mechanics, one can compute the entropy theoretically. Such calculations have been carried out for a great many substances—in particular for carbon tetrachloride (see Table 1 in Chapter 2). These calculations yield results in excellent agreement with those obtained by experiment.

This affords another confirmation of Boltzmann's formula, since the above theoretical calculations are based on it. From the formula (4.16) for the statistical weight of a state of a system made up of atoms and molecules, we derived Boltzmann's distribution (4.20) corresponding to equilibrium. This led in turn to the canonical partition function Z, in terms of which one can compute theoretically the chief thermodynamic quantities of such a system at equilibrium, obtaining theoretical results in agreement with experiment.

However, Boltzmann's distribution is not of universal applicability; it applies to systems consisting of atoms and molecules, but not to those made up of electrons and photons.

The formula (4.16) depends on the assumption that the particles of the system are distinguishable, that they can be numbered, or imagined to be painted different colors. This assumption fails for the microparticles studied in quantum mechanics, since these are indistinguishable in principle. Hence there arises the need to consider different distributions. There are two types of quantum statistics: the Bose-Einstein distribution, used, in particular, in investigations of the so-called photon gas, and the Dirac-Fermi distribution, used, in particular, in investigating electrons—the electron gas in metals, for example. Whereas any number of photons may be at the same energy level, only two electrons may be at the same energy level, and then with opposite spins, that is, characteristic angular momenta.[12] That is why two further statistics were needed in addition to Boltzmann's distribution. However, we shall penetrate no further into the realm of quantum mechanics.

Gibbs' paradox

The formula for the entropy of a mixture derived earlier leads to a remarkable paradox, to which Josiah Willard Gibbs, one of the founding fathers of statistical physics, devoted a series of articles written between 1875 and 1878.

As we showed above (see (4.13)), when N_1 molecules of one gas are mixed with N_2 molecules of another gas, the entropy increases by the amount

$$\Delta S = k\left(N_1 \ln \frac{N_1+N_2}{N_1} + N_2 \ln \frac{N_1+N_2}{N_2}\right).$$

The procedure on which this formula was based was as follows: The two gases were considered to be in a container in unmixed form, separated by a partition. The partition was then removed so that the gases could mix freely.

Suppose now that the numbers of molecules N_1 and N_2 are equal: $N_1 = N_2 = N/2$. In this case the above formula becomes[13]

$$\Delta S = \frac{kN}{2} \cdot 2\ln 2 = kN \ln 2. \tag{4.32}$$

Such is the situation if the gases are different. However formula (4.32) presumably holds even if the gases differ by arbitrarily little: upon being mixed the entropy increases by the amount $kN \ln 2$ regardless. On the other hand, if the gases are identical, that is, if the two halves of the vessel contain one and the same gas, then surely removing the partition cannot cause a change in entropy. We are faced with

[12] According to "Pauli's exclusion principle", in an atom or other system of electrons, there can exist at most one electron with a given set of four quantum numbers giving the energy, and the orbital, spin, and total momenta. *Trans.*

[13] Since $\ln 2 < 0$, this actually represents a decrease in entropy. *Trans.*

a paradox: in a continuous transition from gases differing from one another less and less to gases that are absolutely identical, there is a jump in the behavior of the change in entropy. When the gases differ by ever so little, their mixing entropy remains fixed at $kN \ln 2$. If we imagine this difference to gradually disappear, then in the limit as they become identical, the mixing entropy goes discontinuously from $kN \ln 2$ to zero. What is going on here?

The resolution of this paradox came from quantum mechanics. We imagined a continuous convergence of all the properties of the two gases. (It was also tacitly assumed that if the two gases can be mixed, then their separation can also be effected—of course, at the expense of a certain amount of work.) However, in fact the properties of atoms and molecules are ultimately of a quantum nature, and cannot, therefore, change continuously. For example, there is no continuity of change in the properties of isotopes of an element, since their atomic nuclei differ in the number of neutrons, which of course cannot be other than a whole number. In the same way, there is no continuous gradation between properties of similar chemical compounds, for example, hydrocarbons of the form C_nH_{2n+2}. For large n this compound will differ structurally very little from $C_{n+1}H_{2n+4}$ (where n has been replaced by $n+1$), but n can take only positive integer values, so there is no continuous merging of two substances possible

On the other hand, the Soviet physicists V.L. Lyuboshits and M.I. Podgoretskiĭ have shown that there *are* situations where properties of gases do change continuously from one gas to another. However in such cases the mixing entropy also changes continuously, subject to

$$0 \leq \Delta S \leq kN \ln 2.$$

Unfortunately we cannot go further into these interesting situations in this book, beyond saying that they are of very great interest for theoretical physics.

Nernst's theorem

In Chapter 2 we mentioned in passing the third law of thermodynamics, also known as Nernst's theorem, asserting that at absolute zero entropy is zero. This law is of great importance since it allows us to find not only differences in entropy, but also to assign a meaning to absolute entropy.

From Boltzmann's formula

$$S = k \ln P$$

it follows that S vanishes if the statistical weight $P = 1$. A rigorous proof of Nernst's theorem, which is based on quantum mechanics, uses this observation. A body at absolute zero is in a fundamental state of least energy E_0 of statistical weight 1, an authentic state.

Formula (2.21), giving the dependence of entropy on temperature, namely,

$$S = R \ln V + C_V \ln T + a,$$

is in disagreement with Nernst's theorem: if we let T tend to zero here, then $S \to -\infty$. The fact of the matter is that formula (2.21), which we derived for a classical ideal gas, does not hold at low temperatures—at which, in any case, an ideal gas cannot even exist.

The third law of thermodynamics—Nernst's theorem—has important consequences. In particular, it follows that if the entropy falls to zero, then the heat capacities C_p and C_V must also become zero, as well as the coefficient of thermal expansion.[14]

The third law has been verified by means of direct experiments measuring entropy. All the same, one should be aware of the fact that there are situations where the third law cannot in its strictest sense be valid. Consider, for example, chlorine. In its gaseous state chlorine is a mixture of isotopic molecules, mainly ^{35}Cl—^{35}Cl, ^{37}Cl—^{35}Cl, ^{37}CL—^{37}Cl. Thus the entropy of chlorine gas inevitably includes a mixing component, which remains unchanged no matter what the temperature; in particular, it stays the same at absolute zero. Hence as $T \to 0$ the entropy of chlorine tends to the value of the mixing entropy. In order to change this contribution to the entropy, one would need to alter the isotopic composition of the mixture. A more rigorous investigation would require that one take into account the fact that the energies of the fundamental states of the various isotopic molecules differ slightly from one another, which explains why actual chlorine is not in an equilibrium state at 0°K. In practice, since in ordinary chemical and physical processes the isotopic composition of chlorine does not change, one may assume its entropy to be zero at absolute zero; in calculating only changes in entropy, no errors can arise from this assumption.

Another example is furnished by glass. L. D. Landau once said to me: "What can one say about glass other than that its entropy at absolute zero is non-zero." Actually, a lot can be said about glass, but as far as thermodynamics is concerned, Landau was right.

Glass is a solid substance which is not in equilibrium. When one cools a liquid, then at a certain temperature, when the condition of equality of the free energies of liquid and solid is satisfied, the liquid crystallizes. However, this takes time; the molecules of the liquid must turn and group themselves so as to fit accurately into a crystalline lattice. If the liquid is viscous, this process may take a considerable length of time. When the temperature is lowered, the mobility of the molecules of the liquid decreases, and its viscosity increases. Thus before the crystal manages to form, the molecules in a viscous liquid may have lost nearly all their mobility. This is what happens in the case of glass. For glass to turn into a crystalline solid takes a great deal of time—of the order of a thousand years. One sometimes notices opaque spots on antique glass vessels in museums; these

[14] Measuring the expansion of a body in response to heating. *Trans.*

are where the glass has managed to crystallize. There are several substances that are very difficult to obtain in crystalline form; they prefer, as it were, to "vitrify". Such are polymers, glycerin, and certain mixtures of metallic and flint oxides out of which ordinary glass is made. Thus glass is a hard substance preserving the disordered structure of the original liquid at the temperature at which it vitrified. Here the word "disorder" means that the molecules are mixed up haphazardly in various positions and orientations. Hence it is clear that the entropy of glass at absolute zero cannot be zero; it becomes zero only when the glass crystallizes—and for that to happen we must wait a very long time.

Chapter 5

Statistics and mechanics

Simon Laplace tries yet once more
To calculate everything fate has in store:
When a comet will stream through the heavens,
When the roulette wheel will come up evens,
What new law the king will pass,
Napoleon's return, at last.
But to one thing he didn't tumble:
The opposing player likes to gamble.
So Simon's demon's quit the field—
To sheer chance he's forced to yield.

The distribution of velocities, and temperature

We consider some further consequences of Boltzmann's law.

If an ideal gas is not acted upon by any forces (we neglect gravitational forces, among others), then its molecules possess only kinetic energy, given for each molecule by $mv^2/2$, where m is the molecule's mass and v its speed. The values of v can range from zero to the speed of light. The molecules move in all directions, collide elastically with the walls of the vessel containing them, and with one another, interchanging their speeds.

In 1860, prior to Boltzmann's discovery of his law of the distribution of energies, Maxwell solved the problem of the distribution of speeds of the molecules of an ideal gas, deducing a formula for the proportion of molecules with speeds in the interval from v to $v + \Delta v$. We can now understand Maxwell's distribution as a special case of Boltzmann's distribution where the molecules possess only kinetic energy.

Velocity $\mathbf{v}$ is a vector quantity; it has a direction as well as magnitude, which is the speed $v = |\mathbf{v}|$. Figure 5.1 depicts a rectangular spatial coordinate

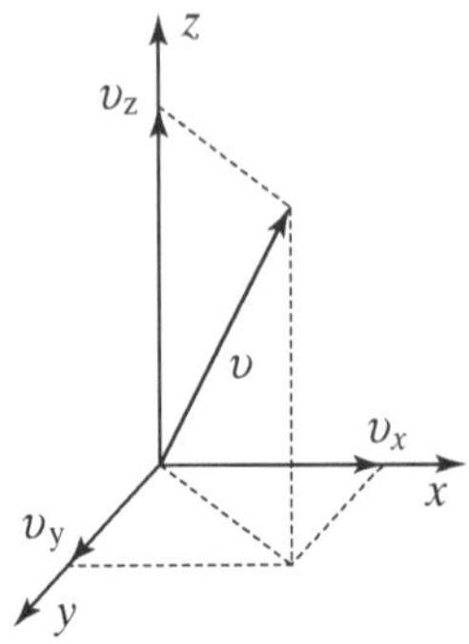

Figure 5.1: A velocity vector and its components.

system, with the axial components v_x, v_y, v_z of a vector $\mathbf{v}$ indicated. It follows from Pythagoras' theorem that

$$v^2 = v_x^2 + v_y^2 + v_z^2. \tag{5.1}$$

We wish to estimate the number of particles with velocities between $\mathbf{v}$ and $\mathbf{v} + \Delta\mathbf{v}$. This is given by the formula

$$dN(\mathbf{v}) = Nf(v)d\mathbf{v}, \tag{5.2}$$

where N is the total number of particles, $N(\mathbf{v})$ is the number having velocity parallel to $\mathbf{v}$ with speed less than or equal to v, and $f(v)$ is the function of the speed we are seeking.[1] According to Boltzmann's distribution (4.20) (or (4.23)), since the energies involved are just the kinetic energies of the molecules, the function $f(v)$ should be given by

$$f(v) = \frac{\exp(-mv^2/2kT)}{Z} = \frac{\exp[(-m/2kT)(v_x^2 + v_y^2 + v_z^2)]}{Z}. \tag{5.3}$$

Since the speeds can take on any values, the value of the kinetic energy varies continuously; in other words, the energy of translational motion—kinetic energy—is not quantized. For this reason the statistical sum Z (see (4.22)) here becomes a statistical integral. And then since the component velocities v_x, v_y, v_z vary independently of one another, we can integrate with respect to each of them separately:

$$Z = \int_{-\infty}^{+\infty} \exp\left(-\tfrac{mv_x^2}{2kT}\right) dv_x \times \int_{-\infty}^{+\infty} \exp\left(-\tfrac{mv_y^2}{2kT}\right) dv_y \times \int_{-\infty}^{+\infty} \exp\left(-\tfrac{mv_z^2}{2kT}\right) dv_z. \tag{5.4}$$

[1]Since $\Delta N(\mathbf{v}) = N(\mathbf{v} + \Delta\mathbf{v}) - N(\mathbf{v})$ is the number of particles with velocities between $\mathbf{v}$ and $\mathbf{v} + \Delta\mathbf{v}$, it follows that $\Delta N(\mathbf{v})/\Delta v$ is the number per unit of speed, and $\frac{\Delta N(\mathbf{v})}{N\Delta v} = f(v)$ is therefore the proportion per unit speed of particles with velocities between $\mathbf{v}$ and $\mathbf{v} + \Delta\mathbf{v}$. *Trans.*

Where do the limits of integration $-\infty$ and $+\infty$ come from? Of course, the speed of a material particle cannot exceed the speed of light; however, for values of the speed v such that, for instance, $mv_x^2 \gg 2kT$, the quantity $\exp(-mv_x^2/2kT)$ is vanishingly small since $e^{-x} \to 0$ rapidly as $x \to \infty$. This being so, we can safely extend the limits of integration to infinity in both directions.

It turns out that these integrals can be evaluated precisely; thus

$$Z_x = \int_{-\infty}^{+\infty} e^{-\frac{mv_x^2}{2kT}} dv_x = \left(\frac{2\pi kT}{m}\right)^{1/2}, \tag{5.5}$$

and similarly for the other two integrals Z_y and Z_z. Hence

$$Z = Z_x Z_y Z_z = \left(\frac{2\pi kT}{m}\right)^{3/2}. \tag{5.6}$$

Substituting for Z in (5.3) and then for $f(v)$ in (5.2), we obtain

$$dN(\mathbf{v}) = N\left(\frac{2\pi kT}{m}\right)^{-\frac{3}{2}} \exp\left(-\frac{mv^2}{2kT}\right) d\mathbf{v}. \tag{5.7}$$

For the components of the velocity we have similarly

$$dN(v_x) = N\frac{\exp(-mv_x^2/2kT)}{Z_x}dv_x = N\exp\left(-\frac{mv_x^2}{2kT}\right)\cdot\left(\frac{2\pi kT}{m}\right)^{1/2} dv_x, \tag{5.8}$$

with similar expressions for $dN(v_y)$ and $dN(v_z)$. We have thus arrived at the Maxwell-Boltzmann velocity distribution. Figure 5.2 shows the graphs of the function $N(v_x)$ for two values of the temperature T. At higher temperatures the graph is wider and flatter. Hence the probability of there being more molecules with high speeds increases with the temperature. Since $N(v_x)$ is an even function the graph is symmetric about the axis $v_x = 0$, at which value of v_x the distribution has its maximum. Hence the average speed is zero. This is to be expected, since the motions of the molecules in the same direction as $\mathbf{v}$ or its opposite are equally likely, so that, on average, at any time the number with velocity component v_x will be approximately the same as the number with component $-v_x$.

Although the average speed is zero, the average of the square of the speed is, of course, different from zero. We shall now compute the average kinetic energy of molecules, and thereby the average square of the speed.

We first find the distribution of the speeds $v = |\mathbf{v}|$. These are non-negative, and may be taken as varying from 0 to ∞. The velocity vector can have any direction, determined by the point coinciding with the tip of $\mathbf{v}$ on the sphere of radius v, centered at the origin (see Figure 5.3). The surface area of this sphere is $4\pi v^2$. Hence, in view of (5.7), the average, or expected, number $N(v)$ of molecules with speed v is

$$N(v) = N(\mathbf{v})\cdot 4\pi v^2 = N\left(\frac{2\pi kT}{m}\right)^{-3/2} \exp\left(-\frac{mv^2}{2kT}\right)\cdot 4\pi v^2. \tag{5.9}$$

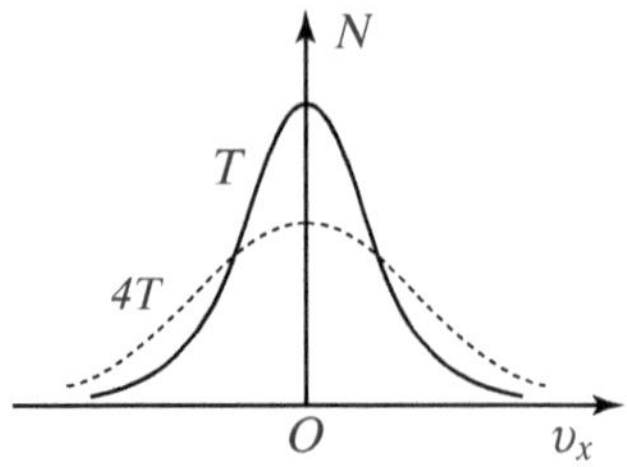

Figure 5.2: Graph of the Maxwell-Boltzmann distribution at temperatures T and $4T$.

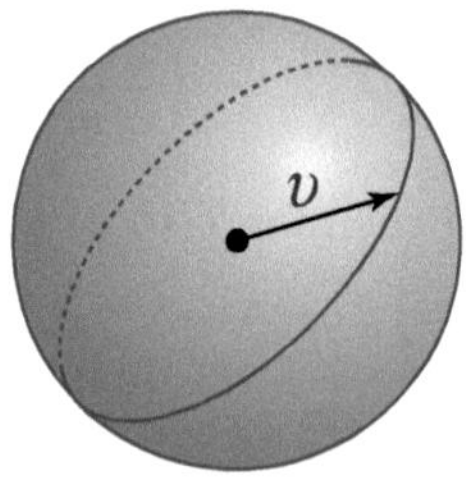

Figure 5.3: This figure relates to the calculation of the mean square speed.

The average kinetic energy of the molecules of the gas can now be computed from this function:

$$\frac{1}{2}m\overline{v^2} = \frac{1}{N}\int_0^\infty \frac{1}{2}mv^2 N(v) \cdot 4\pi v^2 dv, \tag{5.10}$$

which, without too much difficulty, yields

$$\frac{1}{2}m\overline{v^2} = \frac{3}{2}kT. \tag{5.11}$$

Hence

$$\overline{v^2} = \frac{3kT}{m}, \text{ and } \sqrt{\overline{v^2}} = \sqrt{\frac{3kT}{m}},$$

and we have found the mean kinetic energy of the molecules, and their mean quadratic speed.

The formula (5.11) relates the temperature of a monatomic ideal gas directly with the mean kinetic energy of its molecules. Raising the temperature increases the mean kinetic energy proportionately. Thus equation (5.11) affords a kinetic definition of thermodynamic temperature.

We see, therefore, that the crux of the matter is average energy: temperature is directly proportional to *average* energy. It makes no sense to speak of the temperature of a single molecule, nor of its entropy. On the other hand, the *energy* of a single molecule does have a real meaning.

Temperature and entropy measure the average behavior of large collections of particles—they are statistical terms. Pure mechanics is not concerned with temperature or entropy; these are quantities reserved to thermodynamics and statistical physics.

We shall write the Boltzmann distribution (4.20) in the following simplified form:

$$N(E) = NBe^{-E/kT}dE. \tag{5.12}$$

Here $N(E)$ denotes the proportion of molecules with energies between E and $E + dE$, and B is the appropriate pre-exponential factor needed to make this equation identical with (4.20).

Boltzmann's distribution enables us to understand how the speed of a chemical reaction depends on the temperature. As we saw in Chapter 3 (see in particular Figure 3.2), only those molecules react whose energy is sufficient to surmount the "activational barrier", that is, with energy at least equal to the activation energy E_a. By (5.12) the proportion of such molecules is proportional to the quantity

$$e^{-E_a/kT}.$$

An example of such a reaction is the gaseous reaction

$$HCL + NH_3 \rightarrow NH_4Cl,$$

in which hydrochloric acid and ammonia yield ammonium chloride. The speed of the reaction is proportional to the number of collisions of interacting molecules per unit time, that is, to the product of their concentrations, which we denote by means of square brackets:

$$v = \kappa[HCl][NH_3], \tag{5.13}$$

where κ is the proportionality factor between the speed and the concentrations, the so-called *rate constant*. It is this constant that depends on the temperature. On the basis of Boltzmann's distribution, the Swedish physical chemist S. Arrhenius[2] conjectured that

$$\kappa = Ae^{-E_a/kT}, \tag{5.14}$$

where A is a constant.

However, it follows from the discussion in Chapter 3 (see (3.16)) that a more rigorous formula should involve not the energy, but the *free* energy:

$$\kappa = A'e^{-G_a/kT} = A'e^{S_a/k}e^{-H_a/kT}, \tag{5.15}$$

where A' is a constant. In any case, both formulae (5.14) and (5.15) predict that the rate constant κ should depend on the temperature in the same way. By taking logarithms of both sides of (5.14), we obtain

$$\ln\kappa = \ln A - \frac{E_a}{kT}, \tag{5.16}$$

[2]Svante Arrhenius (1859–1927), Swedish physicist and chemist, Nobel laureate in chemistry in 1903. *Trans.*

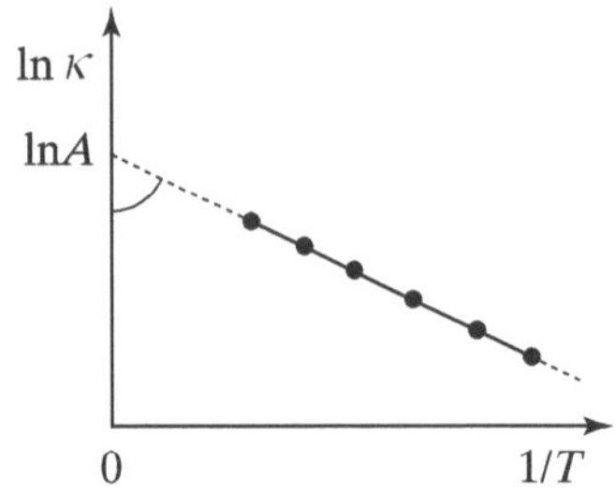

Figure 5.4: Graph of the logarithm of the rate constant against the reciprocal of the temperature, according to Arrhenius' law.

so that $\ln \kappa$ depends linearly on T^{-1} (see Figure 5.4). ((Formula (5.15) implies the same sort of dependence.) By measuring the reaction rates of a reaction at various temperatures, and drawing the graph based on the results, one can infer the activation energy of the reaction from the slope of the (straight-line) graph. All this is well supported by experiment.

The barometric formula and the "gravitational perpetuum mobile"

A special case of the distribution of the molecules of a gas by energy is their distribution in a gravitational field. Consider a thin horizontal layer of gas of thickness dx whose cross-sectional area is $s = 1$ cm^2. The weight of this layer is then

$$nMgsdx,$$

where n is the density, that is, the number of molecules of the gas occupying a volume of 1 cm^3, M is the total mass of the molecules, and g is the acceleration due to gravity near the earth's surface. At equilibrium, this weight must be equal to the force of pressure from below less that from above, that is,

$$-dp = nMgdx. \tag{5.17}$$

(The pressure is the force per unit of cross-sectional area, which is $s = 1$ cm^2.) We shall assume the gas ideal, so that

$$p = nRT,$$

whence

$$dp = RTdn. \tag{5.18}$$

Substituting for dp in (5.17), we obtain

$$\frac{dn}{n} = -\frac{Mg}{RT}dx. \tag{5.19}$$

Integrating this equation from $x = 0$ (the earth's surface) to x, we obtain

$$n = n_0 \exp(-Mgx/RT), \tag{5.20}$$

whence

$$p = p_0 \exp(-Mgx/RT). \tag{5.21}$$

Here n_0 and p_0 are respectively the density and pressure of the gas at sea level. Dividing M and R by Avogadro's constant $N_A = 6.06 \cdot 10^{23}$ per mole, we obtain

$$n = n_0 \exp(-mgx/kT), \tag{5.22}$$

where m is the mass of a single molecule. Formula (5.21) (or (5.22)) is called the *barometric formula*, since it shows how the pressure (or density) of a gas decreases with increasing altitude in the earth's gravitational field. The form of the right-hand side of the formula indicates clearly that this is a special case of Boltzmann's distribution (5.12) when the energy E of a molecule of gas is taken to be its potential energy at altitude x:

$$E = mgx.$$

The barometric formula has rendered important service to physics. It has helped in the determination of Avogadro's number N_A. In 1908, the French physicist Jean Perrin[3] determined experimentally the function

$$f(x) = e^{-mgx/kT}$$

for a colloidal suspension of particles of gamboge[4] in water. For particles of sufficiently large mass m, this function decreases rapidly with increasing height x—by fractions of a millimeter—and one can directly observe the changing values of $f(x)$. Taking logarithms, we have

$$\ln(f(x)) = -\frac{mgx}{kT} = -\frac{N_A mgx}{RT}.$$

Now the mass m of the colloidal particle, the gas constant R, the temperature T, and the acceleration due to gravity g are all known—as is the appropriate correction according to Archimedes' law, needed in view of the fact that the particles are suspended in water. Hence accurate experimental measurements of $f(x)$ could be used to obtain a good estimate of N_A, as was done by Perrin.

In certain publications on popular science, and even in some textbooks, there appeared at one time putative refutations of the barometric formula and the second law of thermodynamics, proposing a "gravitational" perpetual-motion machine of the second kind. It was claimed that, owing to the presence of the gravitational

[3] Jean Baptiste Perrin (1870-1942), French physicist, Nobel laureate in physics for 1926. *Trans.*
[4] A fairly transparent, dark mustard-yellow pigment. *Trans.*

field, in a column of gas at equilibrium there will be a decreasing temperature gradient with increasing altitude x, and that this is achieved without any work needing to be done. This spontaneously arising difference in temperatures could then be used to obtain a practically inexhaustible amount of energy "free of charge".[5]

Here are the details of the argument for a temperature gradient in a vertical column of gas under gravity. Suppose that a certain quantity of gas rises adiabatically, that is, without any exchange of heat, through a vertical distance dx. By (5.17), the pressure will then drop by the amount dp. Now we know that when a gas expands adiabatically its temperature decreases. A simple calculation shows that dT/dx, the temperature gradient (a negative quantity), would then be proportional to the ratio g/C_p, where C_p is the heat capacity of the gas at constant pressure. From the viewpoint of the kinetic theory of gases, this means that the mean kinetic energy of the molecules of the gas decreases with increasing height, part of it having been converted into potential energy ($mgdx$ for each molecule).

This all looks like a very reasonable claim. However it contradicts the barometric formula (5.21) (or (5.22)) just derived. In establishing that formula, we assumed the temperature constant, not varying with the altitude x. Our initial assumption was that the column of gas or colloidal suspension was in a state of thermodynamic equilibrium, so that the temperature would have to be uniform throughout. We also had the barometric law confirmed splendidly by the fact that it has the form of an appropriate special case of Boltzmann's distribution, and by its use in determining Avogadro's number.

What then are the mistakes in the argument for a "gravitational" perpetual-motion machine?

The adiabatic rising of a layer of gas is simply not possible. Gas in a column has no adiabatic chambers, and when a layer rises through a height dx it immediately exchanges heat with the neighboring layers, and the temperature evens out.

The argument concerning the kinetic energy of the gas molecules is also based on an elementary mistake. It is claimed that the kinetic energy of a molecule at height x satisfies

$$\frac{mv^2}{2} = \frac{mv_0^2}{2} - mgx,$$

where v_0 is the speed of the molecule when it was at sea level, and v its speed at height x. However, in reality the slow molecules near the bottom of the column will not have enough kinetic energy to be able to rise to height x.[6] Overall, the

[5] In 1868 Maxwell showed that if there were such a temperature gradient, then a perpetual-motion machine of the second kind would be possible. However both he and Boltzmann argued that at equilibrium the temperature in the column of gas would have to be uniform. The opposing view was held by Loschmidt. In any case, as soon as a temperature gradient is established, would convection begin to occur, disturbing the equilibrium? The inference of irreversible processes (such as stipulated by the second law) from time-symmetric mechanics is sometimes called "Loschmidt's paradox". *Trans.*

[6] Unless knocked upwards by speedier molecules? *Trans.*

only molecules rising to this height are those for which

$$\frac{mv_0^2}{2} \geq mgx.$$

When we compute the mean kinetic energy, the slower molecules make negligible contributions to the result, so that the answer we obtain is the same everywhere in the column. What *docs* decrease with height is the number of particles, that is, the density, and not the mean kinetic energy, that is, the temperature. Precise calculations show this to be indubitably the case.

The reader may object by pointing out that in nature we always observe differing air temperatures at different altitudes. Mount Kilimanjaro is permanently snow-capped, while the climate at its base is tropical. However here there is no thermodynamic equilibrium. The earth is warm and outer space cold. The earth represents an open system heated by the sun (see the next chapter). This could indeed be the basis of an engine, but one completely unrelated to the one we are discussing.

Thus there is no "gravitational" perpetual-motion machine. The appearance of such primitive attempts to disprove the second law is surprising, and their penetration into the literature even more so.[7] On the other hand critique of such attempts has its value.

Physical theory is an integrated, coherent system. If a gravitational perpetual motion machine were shown to be possible, then the barometric formula would have to be pronounced false, along with Boltzmann's distribution, that is, in effect all of thermodynamics and statistical mechanics. Nearly everything written in this book would then turn out to be false. However, as expected, what was in fact easily exposed as false is the possibility of a gravitational *perpetuum mobile.*

Fluctuations

The physical magnitudes used to characterize the states of individual electrons, atoms, and molecules do not in general, of course, coincide with their average values. The behavior of elementary particles is random and subject only to the laws of probability.

At equilibrium the entropy of a system attains its maximum value; the system is in a state of greatest possible disorder. In a given quantity of gas the temperature and density are uniform at equilibrium. The temperature measures the average value of the kinetic energy of the molecules of the gas, and the density gives the average number of molecules per unit volume. However, deviations from the most likely values, from the average behavior, do occur—since, after all, the molecules are in motion, and that motion is random. This is why the entropy can drop

[7]Although one might object that such controversy is good for science. And Loschmidt was no mean scientist. *Trans.*

below the maximum, and temperature gradients and variations in density can spontaneously arise.

These deviations are called *fluctuations*. The greater the number of molecules in the gas the less likely the fluctuations. In Table 4 the statistical weights P of the distributions of N particles in two boxes are tabulated for different values of N using the formula (4.5):

$$P = \frac{N!}{N_1!(N-N_1)!}.$$

Observe how the form of the distribution changes as the number N of molecules increases: the larger N, the more pointed the distribution, that is, the smaller the probability of extreme arrangements of molecules relative to the most probable one, realizable in the greatest number of ways. For $N = 2$ the statistical weight of the state (that is, the number of ways it can be realized) where both molecules are gathered in, say the left box, is half the statistical weight of the state corresponding to uniform distribution of the molecules, that is, one in each box. However for $N = 12$, the ratio of the statistical weights of the corresponding states has gone from $1:2$ to $1:924$. Thus the probability of a significant deviation from the most likely state decreases with increasing N. By the above formula, the statistical weight of the uniform arrangement of N molecules in two boxes is (assuming for simplicity that N is even)

$$P = \frac{N!}{(N/2)!(N/2)!}.$$

From the rough approximation of $N!$ given by (4.10), it follows that

$$P \approx \frac{(N/e)^N}{\left[(N/2e)^{N/2}\right]^2} = 2^N,$$

which is many times larger than the number of ways of realizing any significantly non-uniform, or "ordered" state.

Similar considerations hold for the distribution of other statistical quantities, such as mean kinetic energy of molecules in a gas, or, equivalently, their mean square speed. It follows from equation (5.11) together with the ideal gas equation $pV = RT$, that the mean square speed of molecules of an ideal gas at pressure p is given by

$$pV = \frac{2}{3}N\frac{m\overline{v^2}}{2}, \text{ or } \overline{v^2} = \frac{3pV}{Nm}.$$

For gaseous oxygen one calculates from this that for its molecules O_2, $\sqrt{\left(\overline{v^2}\right)} =$ 455 metres/second. One can calculate from Maxwell's distribution that the probability of finding molecules with speed ten times greater, that is, $v = 4.5$ kilometers/second, is approximately a constant times $e^{-10^2} = e^{-100} \approx 10^{-43}$, a vanishingly small quantity.

Table 4: Distributions of N molecules in two containers and their statistical weights.

Number N of molecules					
2	4	6	8	10	12
$2 \mid 0$ 1	$4 \mid 0$ 1	$6 \mid 0$ 1	$8 \mid 0$ 1	$10 \mid 0$ 1	$12 \mid 0$ 1
$1 \mid 1$ 2	$3 \mid 1$ 4	$5 \mid 1$ 6	$7 \mid 1$ 8	$9 \mid 1$ 10	$11 \mid 1$ 12
——	$2 \mid 2$ 6	$4 \mid 2$ 15	$6 \mid 2$ 28	$8 \mid 2$ 45	$10 \mid 2$ 66
$0 \mid 2$ 1	——	$3 \mid 3$ 20	$5 \mid 3$ 56	$7 \mid 3$ 120	$9 \mid 5$ 220
	$1 \mid 3$ 4	——	$4 \mid 4$ 70	$6 \mid 4$ 210	$8 \mid 4$ 495
	$0 \mid 4$ 1	$2 \mid 4$ 15	——	$5 \mid 5$ 252	$7 \mid 5$ 792
		$1 \mid 5$ 6	$3 \mid 5$ 56	——	$6 \mid 6$ 924
		$0 \mid 6$ 1	$2 \mid 6$ 28	$4 \mid 6$ 210	——
			$1 \mid 7$ 8	$3 \mid 7$ 120	$5 \mid 7$ 792
			$0 \mid 8$ 1	$2 \mid 8$ 45	$4 \mid 8$ 495
				$1 \mid 9$ 10	$3 \mid 9$ 220
				$0 \mid 10$ 1	$2 \mid 10$ 66
					$1 \mid 11$ 12
					$0 \mid 12$ 1

Fluctuational deviations from the mean statistical state of a system are the less likely the larger the deviations and the larger the number of particles in the system. For example, from Table 4 we see that the probability of all molecules being in one of the two containers is very small compared with that of the distribution $6|6$, while the less extreme distributions $7|5$ and $5|7$, or even $8|4$ and $4|8$, are much less improbable.

We shall now calculate the fluctuations in the densities of particles of a system, that is, in the number of particles per unit volume. Suppose we have a quantity of an ideal gas of volume V containing N molecules. Consider a small region of the gas of volume $\nu \ll V$ (Figure 5.5). What is the mean deviation of the number n of molecules in this small region from the mean number $\bar{n}$, that is, the average of the number of molecules over all regions of volume ν? Clearly,

$$\overline{n - \bar{n}} = \bar{n} - \bar{n} = 0,$$

since overall the regions with fewer than $\bar{n}$ molecules will compensate for those with more than $\bar{n}$ molecules. For this reason it is appropriate to consider instead the mean of the *squares* of the quantities $n - \bar{n}$, that is

$$\operatorname{var}(n) = \overline{(n - \bar{n})^2},$$

called the *variance* of the distribution of values n.[8]

[8]In probability and statistics, the variance is one measure of the dispersion of a distribution of values assigned to events in a sample space of a random process. Another is the *standard deviation* σ, the square root of the variance. *Trans.*

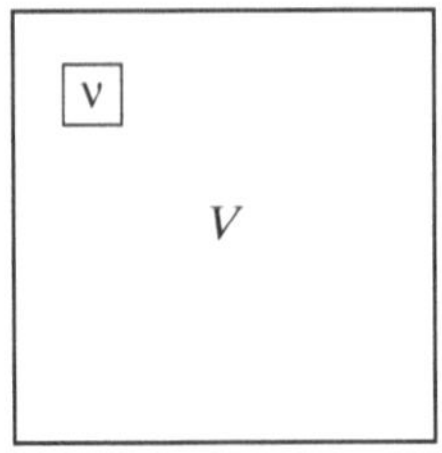

Figure 5.5: This figure relates to the calculation of fluctuations in density.

The probability of finding a particular molecule in a particular region of volume ν is

$$q = \frac{\nu}{V},$$

since it is assumed that the molecules move about in the gas independently of one another. What is the average number of molecules in a region of volume ν? From the simple proportion between

$$(N \text{ molecules occupying } V) \quad \text{and} \quad (\bar{n} \text{ molecules occupying } \nu),$$

we obtain

$$\bar{n} = N\frac{\nu}{V} = Nq.$$

Here is a more rigorous proof of this result. Assume the molecules numbered from 1 to N. We define a quantity Δ_i, where i ranges from 1 to N, as follows: $\Delta_i = 1$ if the ith molecule is in the particular region of volume ν we are considering, otherwise set $\Delta_i = 0$. Then

$$n = \Delta_1 + \Delta_2 + \cdots + \Delta_N = \sum_{i=1}^{N} \Delta_i,$$

and

$$\bar{n} = \bar{\Delta}_1 + \bar{\Delta}_2 + \cdots + \bar{\Delta}_N = \sum_{i=1}^{N} \bar{\Delta}_i.$$

Now what is the value of $\bar{\Delta}_i$? Well, $\Delta_i = 1$ with probability q and $\Delta_i = 0$ with probability $1 - q$, whence

$$\bar{\Delta}_i = 1 \cdot q + 0 \cdot (1 - q) = q.$$

Hence

$$\bar{n} = \sum_{i=1}^{N} q = Nq,$$

and we have arrived at the desired—and in any case obvious—conclusion.

We shall use a similar argument to calculate the mean square deviation from the mean.[9] Observe first that since means are additive,

$$\overline{(n-\bar{n})^2} = \overline{n^2} - 2\bar{n}\bar{n} + \bar{n}^2.$$

Note that, in general, the average of the square of a random variable is different from the square of the average. Here we have

$$\overline{n^2} = \overline{\left(\sum_{i=1}^{N} \Delta_i\right)^2} = \sum_{i=1}^{N} \overline{\Delta_i^2} + 2\sum_{i=1}^{N}\sum_{j>i} \overline{\Delta_i \Delta_j}.$$

We shall now simplify the right-hand side using elementary methods. Observe that

$$\Delta_i^2 = \Delta_i$$

since Δ_i is either 0 or 1. Hence

$$\overline{\Delta_i^2} = \bar{\Delta}_i = q,$$

whence for the first sum on the right-hand side above, we have

$$\sum_{i=1}^{N} \overline{\Delta_i^2} = \sum_{i=1}^{N} \bar{\Delta}_i = \sum_{i=1}^{N} q = Nq.$$

Now we turn to the double sum. What is $\overline{\Delta_i \Delta_j}$? We have

$$\overline{\Delta_i \Delta_j} = 1 \cdot 1 \cdot q^2 + 2 \cdot 1 \cdot 0 \cdot q(1-q) + 0 \cdot 0 \cdot (1-q)^2 = q^2.$$

Hence

$$\overline{n^2} = Nq + 2\frac{N(N-1)}{2} q^2,$$

since there are $N(N-1)/2$ pairs (i,j) with $i < j$. Hence, finally,

$$\begin{aligned}\overline{(n-\bar{n})^2} = \overline{n^2} - \bar{n}^2 &= Nq + N(N-1)q^2 - N^2q^2 \\ &= Nq(1-q) = \bar{n}(1-q).\end{aligned}$$

Since $q \ll 1$, we can neglect q compared with 1, whence

$$\overline{(n-\bar{n})^2} \approx \bar{n}. \tag{5.23}$$

Our final result looks very surprising. It is of great significance.

A fluctuation represents a change in the quantity P, the statistical weight of the state. The equilibrium state corresponds to the maximal statistical weight P_m, while a fluctuation from this state will correspond to a smaller statistical weight

[9] That is, the variance. *Trans.*

P'. The greater the fluctuational deviation is from equilibrium, the less the value of P'. We may interpret the ratio P'/P_m as the probability of a fluctuation to a state with statistical weight P'. From Boltzmann's formula in exponential form (see (4.15))

$$P = e^{S/k},$$

it follows that

$$w = \frac{P'}{P_m} = \exp \frac{S' - S_{max}}{k}. \tag{5.24}$$

Since for any appreciable fluctuation we must have $P' \ll P_m$ and $S_{max} \gg S'$, the probability w of such fluctuations is very small.

The formula (5.24) was deduced by Einstein in 1906.

Why is the sky blue?

Although the probability of fluctuations is small, we encounter them every day. If there were no fluctuations in the density of the air, that is, no fluctuations in the number of particles in small regions, we would see stars in the daytime and the sun would shine in a black sky. In other words, our daytime view of the heavens would be just like that of astronauts. As a rule, we give little thought to the blueness radiating from the firmament. But there is something needing explanation here. Why is it that when we stand with our backs to the sun we see a blue sky? After all, sunlight propagates in a straight line from the sun. It follows that something is being illuminated by the sun. It can only be the air.

S.I. Vavilov[10] regarded Leonhard Euler's *Letters to a German princess on various topics of physics and philosophy*[11] as the best popular book about physics. Letter XXXII of the first volume of this work, dated July 23, 1760, is entitled "On the blueness of the sky". Euler wrote: "I shall demonstrate to your highness that the cause of the blueness of the sky must be sought in our atmosphere, which is not perfectly transparent.... The air is made up of a multitude of small particles, which are not entirely transparent, and, when illuminated by light rays, acquire an oscillatory motion giving rise in turn to new rays characteristic of these particles...these particles are of a bluish color."

These are weighty words. Modern physicists are also of the opinion that illumination excites changes in the state of electrons and molecules, and such changes might be interpreted as these particles' own oscillations. However, Euler was wrong. The molecules of nitrogen, oxygen, argon, and carbon dioxide that form the atmosphere are colorless; they do not absorb light in the visible part of the spectrum.

---

[10]Sergeĭ Ivanovich Vavilov (1891–1951), Soviet physicist. *Trans.*

[11]*Lettres à une princesse d'Allemagne sur divers sujets de physique et de philosophie.* St. Petersburg, 1768–1772, Vols. 1–3; Opera III-11, 12.

More than 100 years later, a new explanation was proposed by John Rayleigh,[12] the English physicist who achieved a great deal in optics and acoustics.

The sky is blue because the air disperses the sunlight in much the same way as light is scattered in various directions different from that of the incoming beam by any semi-opaque or misty medium. Rayleigh's view was that sunlight is dispersed not by relatively large atmospheric inhomogeneities (such as water droplets in mist) but by the air's molecules themselves through their random movement.[13] He produced a formula, now famous, for the intensity of the dispersed light, according to which that intensity is inversely proportional to λ^4, where λ is the wavelength of the incident light. It follows that light rays in the violet region of the spectrum, with wavelengths around 400 nanometers, are dispersed about 13 times more intensely than red light rays, with wavelengths around 760 nanometers. And that is why the sky is blue. That's also why distant forests and mountains appear bluish: we see them through a thick layer of air lit mainly from above and scattering preponderantly the blue light falling on it. This also explains why the sun looks red at sunrise and sunset: at those times there is a thicker region of the atmosphere between us and the sun than when the sun is at its zenith, and what reaches our eyes directly are the rays least dispersed.

Later it was shown that the mechanism of the dispersion of light by air (and other gases) consists in deviations from the homogeneous distribution of the molecules by volume as a result of their continual random motion. Fluctuations in the density of the gas occur, and one can say that the light is dispersed by these fluctuations. The intensity of the dispersed light, which by Rayleigh's law is proportional to λ^{-4}, turns out to also be proportional to the mean square fluctuation $\overline{(n-\bar{n})^2}$ of the number of molecules per unit volume. By (5.23) this is very close to $\bar{n}$, so by measuring the intensity of the dispersion one can estimate the number of molecules per unit volume of the gas, and thence the number in a mole, Avogadro's constant. This has been done to high accuracy.

Following the work of Rayleigh and Einstein, the theory of the dispersion of light in gases, liquids, and crystals was further developed by the Polish physicist M. Smoluchowski and the Soviet physicist L.I. Mandel′shtam.

Thus when we look at the sky, we are in essence seeing fluctuations of the density of the air. Fluctuations in physical quantities limit the sensitivity of physical instruments, so that, for example, measurements of current or voltage fluctuate about some mean value. Fluctuations are responsible for the "noise" in radio-technology and elsewhere.

In Chapter 4, we considered the rolling up of a polymer chain in a statistical ball. A polymer chain is a macromolecule that continually fluctuates in solution,

[12] John William Strutt, 3rd Baron Rayleigh (1842–1919), Nobel laureate in physics, 1904. *Trans.*

[13] Or is it that although sunlight *is* dispersed by large-scale inhomogeneities such as water droplets in mist, this dispersion is not wave-length dependent, whereas the dispersion due to the molecules' movement is smaller in scale and wavelength-dependent. *Trans.*

with the balls lengthening and shortening as a result of heat-induced motion in the form of rotations about single bonds. Statistical theory has been applied to estimate the distribution of the sizes of polymer balls and the average size of a polymer ball.

It is clear that in a system in equilibrium, any fluctuation that appears will after some time disappear—be smoothed out—since the system tends to its most likely state. But then new fluctuations arise. The motions of atoms and molecules, and electrons in metals, for instance, are subject to the laws of chance.

The age of Darwin

Boltzmann was once asked how he would characterize the century he lived in, the 19th. His answer was "The age of Darwin".

This is a profound reply. The laws of chance triumphed earlier in biology than in physics. It was the biologist Darwin who discovered the fundamental mechanism of evolution: variation plus natural selection. Every population of animals or plants undergoes changes—spontaneous, hereditary, or otherwise—, random deviations from the most likely values of the various characteristics of the organism. Consider, for example, a sufficiently large, randomly chosen group of people. In this group height will obey the same law as the distribution of speeds of molecules in a gas. As height h moves away from the average, the number with that height decreases from the average height symmetrically, that is, independently of whether h is less than or greater than the average height (Figure 5.6).

Unlike fluctuational deviations from the mean in the behavior of molecules in a gas, evolutionary variations (whatever the causes) do not die out if the new characteristics are adaptive, that is, ensure better conditions of existence and procreation for the organism. Variability itself changes, evolves. Living organisms develop and retain the imprint of their evolutionary development. We now know that evolution has been going on for about $3.9 \cdot 10^9$ years. Thus evolution is a very different process from that in isolated physical systems of the sort we have been considering thus far, which reach a state of equilibrium relatively quickly.

Nevertheless, the study of heat phenomena and the concomitant discovery of the irreversibility of physical processes allow us to understand the nature of physical evolution. A physical system also evolves—in the direction of maximum entropy, if it is isolated. How *open* systems behave, that is, systems exchanging matter and energy with the surrounding world, will be the theme of the next chapter.

An isolated system is a relatively unexciting affair. Sooner or later it achieves a state of equilibrium at which its entropy is greatest, and thereafter there occur only small fluctuations in the direction of decreasing entropy. One might say that in an isolated system the "arrow of time" is oriented towards maximum entropy. Once equilibrium has been attained, time ceases, as it were—no further events occur, and if we neglect the small fluctuations, there is no further change. However, as

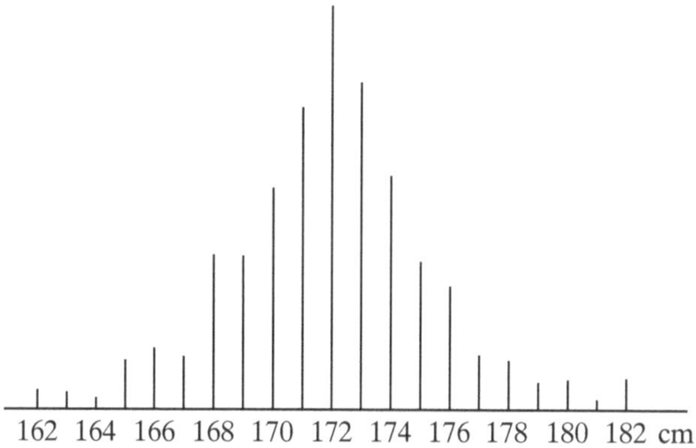

Figure 5.6: Distribution of heights in a large, arbitrarily chosen group of people.

long as equilibrium has not been reached, an isolated system will evolve, "aspiring" to maximize entropy.

Boltzmann set himself the task of creating an evolutionary physics, following on Darwin's creation of evolutionary biology.

What preceded this development? If the 19th century was the age of Darwin, then the 18th was the age of Newton—of the triumph of classical mechanics, the pure mechanics of reversible non-dissipative processes, that is, processes not dissipating energy. In that ideal mechanics there was no place for randomness, no place for entropy. Given the laws of motion, and the initial positions and velocities of the bodies comprising the system, the subsequent behavior of the system was completely determined, and could in principle be calculated in advance. The system "remembered" its initial conditions.

It seemed that everything was solved. In Newton's lifetime, the English poet Alexander Pope could write:

Nature and nature's laws lay hid in night;
God said, let Newton be! And all was light.

The world seemed to be a grandiose mechanism functioning like clockwork, so that once set going it unfolded automatically according to simple rules. In this sense, classical mechanics, in which irreversibility and randomness had no place, was a science of a simple, passive world. But already in 1802, the English poet and artist William Blake wrote:

... May God us keep
From single vision and Newton's sleep!

Although the age of Darwin had not really begun, to this great poet and subtle thinker it had already become clear that classical mechanics was narrow and one-sided.

And indeed now we understand that classical mechanics has definite limits to its applicability. Within these limits it works splendidly: it allows us to compute with great precision the trajectories of the planets and satellites, and to solve a wide range of technological problems. However, the universe as a whole, as an evolving totality of mutually interacting bodies, resisted being encapsulated in a general theory characterized by time-reversibility.

In 1811 the French mathematician and physicist Jean Baptiste Fourier established the law of heat conduction, relating to the unidirectional transfer of heat from a hotter to a colder one. This marked the beginning of the expansion of physics beyond the limits of Newton's scheme. There soon followed Carnot's "Reflections", the discovery of the second law of thermodynamics, and its probabilistic treatment by Boltzmann and Gibbs.

A purely mechanical system "remembers" its initial conditions. In contrast, an increase in entropy of a system signifies a gradual "forgetting" of its earlier more ordered states. Who knows how the liquid in a vessel reached its present temperature? Was it through cooling or being heated? What is important is only the fact of the liquid's being in a state of equilibrium.

Thus in the age of Darwin, the basic problem before Boltzmann was that of deriving the law of increasing entropy from the laws of mechanics. After all, if the thermal properties of bodies are determined ultimately by the motions of their molecules, then there must be a path from mechanics to a theory of heat. The molecules of a gas are in motion just like ordinary macroscopic bodies,[14] with their individual trajectories, that is, with determinate positions and velocities at each instant of time. What does it matter that there are huge numbers of them? Mechanics must apply regardless.

By considering the interactions of molecules in a gas, Boltzmann was able to construct a function H of the coordinates and velocities of the molecules, and to show that it has a bias in its variation with time: $dH/dt \leq 0$. Entropy is proportional to this function.[15] This is the celebrated H-theorem of Boltzmann, of great importance for physical kinetics.

However, Boltzmann failed to solve the problem as he had proposed it. In the derivation of his result he made certain tacit assumptions of a non-mechanical, probabilistic character, namely assumptions about the chaotic behavior of molecules.

The strenuous task of bridging the abyss between mechanics and statistics, and the failure of his contempories to appreciate his work led to Boltzmann's suicide in 1906.

---

[14] Leaving aside considerations of quantum mechanics for the moment. *Trans.*

[15] It turns out that $S = -kH + S_0$, where S_0 is constant, whence it follows that $dS/dT \geq 0$, Boltzmann's "generalized second law of thermodynamics": entropy increases unless the system is in equilibrium. *Trans.*

Laplace's demon and Sinaĭ's billiard[16]

The French astronomer, physicist, and mathematician Pierre Simon Laplace was one of a pleiad of brilliant scientists who continued the work of Newton. Thus he is considered one of the creators of theoretical mechanics. In the course of his work he invented a "demon". A demon also crossed Maxwell's path, but that was a different one, and we shall discuss him in Chapter 7.

By Laplace's demon we mean an imaginary creature of his invention. Laplace was a devoted believer in mechanics. Of course, he knew that an unexceptional quantity of gas contains an exceptionally large number of molecules, which are moving in all directions and colliding with one another and with the walls of the vessel containing them. As we have seen, one can describe the properties of a gas by resorting to average characteristics of its molecules; for instance temperature arises in this way via equation (5.11):

$$\frac{1}{2}m\overline{v^2} = \frac{3}{2}kT.$$

It is impossible to keep track of the motions of each and every molecule of the $2.7 \cdot 10^{19}$ molecules in a cubic centimeter of a gas at atmospheric pressure. However, Laplace was convinced that in fact the motion of each molecule is fully predetermined, and that there is no element of chance involved. If we need to resort to the statistics pertaining to random quantities, that can only be due to our ignorance of the precise facts.

If there existed a demon capable of ascertaining at some initial time t_0 the initial positions and velocities of every molecule of a gas, then he could predict the state of every molecule of that gas at every later time, and therefore what would happen to the gas. Furthermore, if this is so for a cubic centimeter of gas, then it should be true for anything whatsoever. Hence, chance has no objective reality; events in a system that appear to be random, actually hide fully determinable parameters of the system. Although we don't know them, Laplace's demon does.

Laplace wrote: "...we must view the present state of the universe as the effect of its previous state and the cause of the next. A mind that knew at any particular given moment all forces arising in nature, and the relative positions of all of its component parts, and if in addition it were comprehensive enough to subject these data to analysis, would be able to embrace in a single formula the motions of the largest bodies in the universe as well as those of the lightest atoms, and there would be nothing left to conjecture, the future no less than the past would be accessible to its gaze."[17]

How will a tossed coin fall? Heads or tails? This will depend on how it was tossed and on air resistance, that is, on collisions with air molecules. Laplace's

[16] A *billiard* is a dynamical system abstracted from the game of billiards, that is, where a particle alternates between frictionless motion in a straight line and elastic mirror reflection off the boundary. *Trans.*

[17] *Essai philosophique sur les probabilités*, Paris, 1814.

demon can calculate in advance and know for sure; he knows everything, including the future and the past of the world, for the universe evolves according to completely fatalistic laws.

Of course, Laplace knew perfectly well that his demon was an impossibility. The point of the above excerpt from his *Essai* was merely to emphasize that everything in the universe is predetermined, predestined, even though this determinateness is inaccessible to us. In practice, we do not know what lies in store, and cannot discover it.

From this point of view entropy can be taken to be a measure of our ignorance, of the inadequacy of the information at our disposal about the system under investigation. If Laplace is right, then this ignorance is merely subjective, not objective; it does not arise from actual laws of nature independent of our consciousness.

Thus do we see that the problem of the relation between mechanics and statistics leads us to deep philosophical questions, to questions of existence and consciousness.

The question of the relation between mechanics and statistics has been investigated by several Soviet scientists: N.S. Krylov (1948), who died young, the physicist and mathematician N.N. Bogolyubov (1946), and the mathematicians A.N. Kolmogorov (1958) and Ya.G. Sinaĭ (1963, 1967).

Krylov's[18] basic idea was that not every solution of a mechanical problem corresponds to actual experience. Solutions of such problems can be stable or unstable. Consider, for example, a physical pendulum, that is, a rigid body that can oscillate about an axis to which its end is fixed (Figure 5.7). The pendulum has two stationary equilibrium positions, hanging vertically downwards and standing vertically upright. The lower position is stable: a small perturbation from that position will cause the pendulum to swing a few times, gradually returning to the initial position. The upper position is unstable: the slightest touch will be enough

[18]Nikolaĭ Sergeevich Krylov (1917–1947), Soviet theoretical physicist. He arrived at the conclusion that statistical physics cannot be constructed on the basis of classical mechanics—nor on that of quantum mechanics. The following excerpt from the review by E.T. Jaynes (*J. Amer. Statistical Assoc.* **76** (1981)) of the English edition of Krylov's incomplete and posthumously published book *Works on the foundations of statistical physics* may help with the passage that follows. "Since the beginnings of statistical mechanics, two different streams of thought have been competing for...foundation status. The 'ergodic' view associated with James Clerk Maxwell, sees it as a part of mechanics, the goal being to deduce the probability distributions for systems of many molecules by application...of Newton's laws of motion. A very different view, associated with J. Willard Gibbs, sees the goal merely as making the best predictions possible of observed facts, from incomplete information; that is, it is just a branch of statistical inference.... With this background, we can explain that Krylov's work is a deeply thoughtful...statement of the ergodic approach.... Krylov and his followers regard probabilities as in principle determined, as are velocities and accelerations, by the laws of mechanics (but in a more complicated way, not yet fully carried out). They view...[Gibbs'] method as nonrigorous and provisional, since its distributions are not derived from the laws of mechanics. Advocates of the [Gibbsian] view see the ergodic program as not only unnecessary but in principle impossible, because the necessary information is lacking. That is, Newton's laws of motion can tell us only how probabilities change with time, and not what probabilities should be assigned at the start." *Trans.*

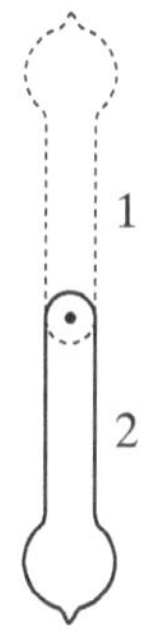

Figure 5.7: Stationary equilibrium positions of a physical pendulum.

to topple the pendulum.

Clearly, in situations of radical instability such as this, the trajectory of each component of the system has no precise meaning. One can observe such trajectories (for example, the Brownian paths of colloidal particles in suspension can be observed under the microscope), but computation of these trajectories from the initial conditions and the laws of motion will never yield a result consonant with actual experience. An unstable trajectory "forgets" its initial conditions.

If a solution of some problem of mechanics represents an unstable equilibrium, then it can never actually be observed in reality. The only solutions that *do* correspond to real situations are those representing stable equilibria. It follows that if a mechanical process is in a state of unstable equilibrium, then it can become irreversible in time.[19] This is related to the limitations of the notion of an isolated system. Clearly, this notion is an idealized one: in reality there are no absolutely isolated systems. However, if the external influences are small and the system's reactions to them are also small, then we may consider the system as approaching an ideally isolated one. On the other hand, if the system is in a state of unstable equilibrium, then a small external action on it will radically alter its state. Obviously such a system cannot be considered even approximately isolated, since minute causes can have large effects. Furthermore, the only method available for investigating such small perturbations is via statistics based on the assumption of their randomness.

An unstable mechanical system—understood as consisting of a vast number of molecules moving in disordered fashion in the sense that their trajectories are unstable—represents the source of the essential statistical nature of such a system. It is not possible to reduce the behavior of a system in unstable equilibrium to the behaviors of its constituent particles by means of mechanics alone. We are compelled to introduce probabilistic considerations, involving the probability of one or another kind of deviation in the motions of the particles.

[19]In the sense that it could never be observed to have returned to the state of unstable equilibrium? *Trans.*

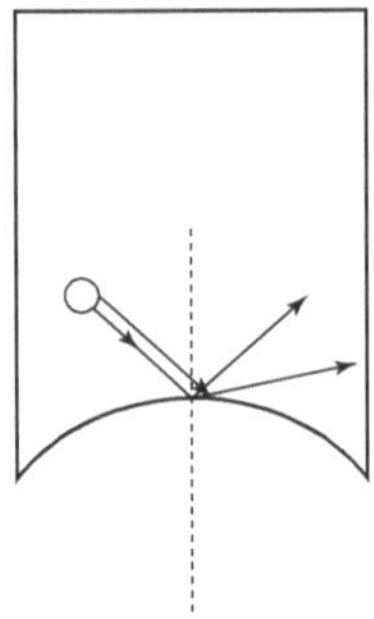

Figure 5.8: Reflection of an elastic particle from a convex wall of the containing vessel.

A probabilistic description of the behavior of a system is entirely objective. It expresses not our ignorance but objectively existing instabilities, that is, randomness of behavior that really does correspond to reality.

And irreversibility arises precisely as a consequence of the impossibility of obtaining a complete description of such behavior in a mechanical system. In general, of course, the incompleteness of the description may arise for various reasons, such as inaccuracy of the initial or boundary conditions, or changes in the motions of the particles caused by the appearance of random forces.

One can show how instability arises and develops using the rigorous language of mathematics. Consider elastically colliding balls in a container. As a 2-dimensional model, we can take a billiard table without pockets on which the billiard balls roll without friction. We assume to begin with that we have calculated the trajectory of a particular ball exactly. With each collision the actual trajectory will deviate by ever so small an angle from the calculated one. However, this deviation grows as more and more collisions occur, and computation shows that it actually increases exponentially with the number of collisions, and therefore with time. In this way there occurs a "forgetting" of the initial conditions, and of the initial trajectory.

It is important to note that this thought experiment does not require many balls. In a version due to Sinaĭ, [20]with just one of its edges convex (Figure 5.8), a single ball suffices. Every trajectory of the ball involving reflection off the convex edge is unstable. Two initially arbitrarily close trajectories may diverge unboundedly.[21]

Thus Laplace's determinacy is impossible. Any attempt to foretell the future starting from initial conditions and the laws of motion—even for such a simple system as that of a single ball in a container with a convex portion of wall—would very soon encounter instabilities and in the end collapse completely.[22]

[20]Yakov Grigor′evich Sinaĭ (1935-), Soviet/American mathematician. *Trans.*

[21]Shades of chaos theory. *Trans.*

[22]But couldn't Laplace retort that the instability here results from inadequate knowledge of the

We find ourselves in a peculiar situation. In mechanics, the idea of the trajectory of a body has real meaning. It is a well-defined concept. On the other hand, quantities defined in terms of statistics and probabilistic averages have no constructive content in mechanics—it is meaningless to talk of the temperature or entropy of a mechanical system. Purely mechanical systems are adiabatic; their entropy does not change: $\Delta S = 0$. *Mutatis mutandis*, in statistical mechanics the concept of the trajectory of a body loses its sense, but perfectly well-defined statistical properties of a system arise—its temperature and entropy.

Energy, however, retains its meaning in both classical and statistical mechanics.

Thus the entropy of a system is indeed an objective measure of our ignorance, of our lack of complete information[23] concerning the system—but this is ignorance *in principle*. Entropy is a measure of the degree of impossibility of knowledge—in principle—where instabilities of trajectories and other mechanical features of the system arise. The absence of information is a property of the system and not of the observer. After all, as we saw at the very beginning, the entropy of a system is a function of its state.

The fate of the universe[24]

Is the world bound to end in heat death? This was Thomson and Clausius's gloomy prognostication when they discovered the law of increasing entropy in irreversible processes. What is most astonishing is that the investigation of a technological process—the working of a steam engine—should lead to such a general cosmological conclusion.

But what is meant by "heat death"?

This is generally held to mean a state of everything, of the whole universe, of maximum entropy, a state characterized by a homogeneous distribution of matter at uniform temperature. Thinking of the universe as an isolated system, we must conclude that this temperature must be absolute zero. Everything grows cold, freezes, and evolves into a dead chaos. According to the second law, in an isolated system in such a state—the most likely one—there can arise no inhomogeneities,

---

exact initial and boundary conditions? Wasn't his argument about what is possible *in principle*, rather than in practice? *Trans.*

[23] The technical definition of "information" will be given in Chapter 7.

[24] Since the original of this book was published in 1986, there have been a great many new developments in physical cosmology. For example, the age of the universe is now thought to be around 14 billion years, rather than the 20 billion mentioned by the author. However, although there are several theories as to the evolution of the universe since the big bang, including cyclic ones, the idea that soon after that event the expanding universe was in a state of very, even somewhat miraculously, low entropy, and that entropy has been increasing ever since continues to hold sway. Brian Greene's book *The fabric of the cosmos* (Penguin Books, 2005) is an excellent, relatively recent, popular account by an expert. *Trans.*

no order, no information.[25]

Thomson, who was apparently unaware of statistical mechanics and the probabilistic interpretation of entropy, viewed the heat death of the universe as inevitable.

But Boltzmann, the discoverer of the connection between entropy and statistical weight, believed that the present, obviously highly inhomogeneous state of the universe with its stars and galaxies, came about as the result of a large-scale fluctuation.[26] If the universe had ever in the past been homogeneous and isothermal, then only a fluctuation could account for its present inhomogeneity—an immense fluctuation and therefore of infinitesimally small probability. But over the span of cosmic time, measured in billions of years, anything might happen....

However, while these arguments were very convincing at the time of their formulation, subsequent discoveries concerning the structure and history of the universe have called them into question.

The universe is definitely not at present in a state of statistical equilibrium. This is also the case for any arbitrarily large finite region of the universe. Boltzmann looked to fluctuations for an explanation of this remove from equilibrium. However, it is on the basis of the general theory of relativity that the true explanation can be arrived at. A gravitational field, responsible for the creation of the stars and galaxies, permeates the universe. According to the general theory of relativity, this field causes changes in space and time. In ordinary, non-relativistic, statistical physics, in terms of which the heat death hypothesis was formulated, space and time were regarded as external factors, a fixed backdrop. The second law of thermodynamics is valid only under stationary external conditions. However the universe turned out not to be stationary, but expanding. It is a system immersed in a changing gravitational field.

That gravitational forces are capable of producing structure from chaos, that they can cause stars to form from cosmic dust, was known long before the general theory of relativity came into being—Newton had already discussed this. In 1886, the great Norwegian dramatist Henrik Ibsen wrote in his poem "Stars in nebula":

A message to our ancient earth it[27] bore,
That in the solemn stillness of the distance
There chaos had evolved a star-existence

[25] See Chapter 7 for the precise definition of "information".

[26] Boltzmann realized that, since the laws of classical physics (and now also the standard version of quantum theory) are time-reversible, and, presumably, the behavior of systems consisting of many particles is ultimately subject to those laws, one might equally logically conclude that entropy was *higher* in the past, that is, that entropy increases as t decreases. Since everyday experience seems to contradict this, he took refuge in the hypothesis that the present state of the universe resulted from a huge, and extremely improbable, fluctuation. However, this hypothesis leads to logical impasses—for instance, we cannot then trust our memories or other evidence of a past. The presently favored way out of this difficulty is to hypothesize that very soon after the big bang the universe was in a state of very low entropy, which has been increasing ever since. For further details see Brian Greene's *The fabric of the cosmos* (Penguin Books, 2005). *Trans.*

[27] A comet.

Once it discovered gravitation's law.

Here is the last stanza of the poem:

The nebula, though formless, I believe in,
Chaotic though the weltering North may be;
Believe it is on course for unity,—
A brilliant star in process of conceiving.
(Translated by John Northam.)

The entropy of the universe would seem to be increasing. According to present-day notions, the universe we know came into being some 20 billion years ago as a clot of hot, structureless plasma. Structure began to emerge immediately after the "big bang". First there came into being elementary particles with nonzero rest-mass, later atoms were formed. Eventually gravitational forces caused the stars and galaxies to form. This clumping of atoms and molecules was accompanied by a rise in temperature in the clumps through increasing pressure, and dispersion of heat via radiation, causing the entropy of the surrounding medium[28] to increase. The universe's entropy consists largely of radiation entropy. The *specific entropy* of the universe is expressed by the ratio of the number of photons to the number of heavy elementary particles—primarily protons—per unit volume. This ratio characterizes the degree to which the universe is heated. The photons present in the universe are predominantly surviving relics of the radiation emitted at the moment of the big bang. This radiation has since then cooled to the very low temperature of 3°K,[29] and the number of photons per cubic centimeter is about 500. The average density of matter in the universe is about 10^{-30} grams per cubic centimeter,[30] or about one proton per cubic meter. Hence the ratio of the number of photons to the number of protons is of the order of 10^9. Thus the specific entropy of the universe is very large.

The ultimate fate of the universe depends on a number of phenomena that have not yet been adequately investigated. We mentioned earlier (in Chapter 4) the neutrino, an elusive elementary particle discovered by Pauli theoretically, and only much later actually observed. Until recently it was thought that, like photons, the neutrino has zero rest-mass, but now strong experimental evidence has emerged showing that it may after all have a tiny rest-mass. If this is confirmed, then our ideas of the cosmos will have to be changed, since it is estimated that neutrinos are approximately as abundant as photons. If they do indeed possess rest-mass, then the average density of matter in the universe will have to be drastically revised upwards.[31] The universe can expand forever—be "open"—only if the density of

[28] The atmospheres of the stars? *Trans.*

[29] This is the "cosmic microwave background radiation", discovered by Arno Penzias and Robert Wilson in 1964. *Trans.*

[30] Does this include the mysterious "dark matter" hypothesized to explain why the galaxies cohere? *Trans.*

[31] However, according to *The fabric of the cosmos* (see footnote 26) p.433, neutrinos "are too light to supply the dark matter; they fall short of the mark by a factor of more than a hundred." *Trans.*

matter is less than a certain critical value. If this is exceeded then the "closed" model of the universe applies: after space expands to a certain finite size, it will begin to contract in upon itself.[32]

Another phenomenon connected with the fate of the universe is that of the instability of protons. Certain theoretical considerations related to "grand unification" led to the hypothesis of proton decay, with a half-life of at least 10^{32} years.[33]

In the universe there are continuing processes of restructuring, accompanied by the export of entropy from collapsing regions and an overall increase in entropy. However, increase of entropy in a non-stationary system located in a changing gravitational field, does not at all imply that the system will tend to a homogeneous state and statistical equilibrium.[34]

The point here is that when solving problems concerning entropy, we usually work with the model of an ideal gas, where it is assumed there is no interaction between the molecules beyond elastic collisions. However in the universe gravitation acts, and so has to be incorporated into the picture.

The claim of thermodynamics that a closed system will sooner or later reach a state of maximum entropy is valid only if the conditions external to the system are stationary, or can be ignored. However this is not the case for the universe since the metrical, spatial properties[35] of the expanding universe are time-dependent. Thus statistical equilibrium need not occur, that is, the general theory of relativity may save us from the conclusion that the universe will end in "heat death".

Not every one agrees with this argument. In his book *The big bang*,[36] J. Silk argues that heat death is inevitable for an open universe. He writes: "Although an open universe seems the favored alternative at present, an open universe has a distinctly unappealing future. In an open universe, galaxies are destined to run down, and stars are destined to burn out, never to be reborn. Gravity can no longer counteract the expansion, and gravitational forces become insignificant on the very largest scales. Space will become blacker and blacker. The void between the galaxy clusters will deepen immeasurably. As nuclear energy supplies dwindle, matter can no longer support itself against gravity in gravitationally bound systems. Galaxies, and ultimately even the great clusters, will collapse to form gigantic black holes. Eventually, all matter will become utterly cold, attaining a temperature of absolute

---

[32] It was discovered in 1998 that the universe, or rather the space containing it, is at present undergoing *accelerated* expansion. (See *The fabric of the cosmos*, p. 411.) *Trans.*

[33] "After years of painstaking search for such proton decay in elaborate underground experiments, none was found." (*The fabric of the cosmos*, pp. 267-268.) *Trans.*

[34] "...by the time the universe was a couple of minutes old, it was filled with a nearly uniform hot gas composed of roughly 75 percent hydrogen, 23 percent helium, and small amounts of deuterium and lithium. The essential point is that this gas filling the universe had extraordinarily *low* entropy. ... For huge volumes of gas, when gravity matters, atoms and molecules evolve from a smooth, evenly spread configuration, into one involving larger and denser clumps. (*The fabric of the cosmos*, pp. 171-172.) *Trans.*

[35] Its geometry? *Trans.*

[36] Joseph Silk, *The big bang; the creation and evolution of the universe*, W. H. Freeman, San Francisco, 1980.

zero. All forces will fade and disappear, until a state is reached where nothing will ever change again. Space is infinite, and a cold, black, immutable future is inevitably destined to be attained throughout space. This fate will not occur for billions and billions of years, but it is nevertheless inevitable in an ever-expanding universe."

We have included this lengthy quote to show that there are modern cosmologists who agree with Thomson. Nonetheless, as we have argued above, this gloomy prognostication does not in fact follow from the general theory of relativity.

Incidentally, the thermodynamic properties of black holes are of great interest, and are in fact very different from those proposed in Silk's book. However, we shall not go into this matter here.

Thus an open universe does not at all tend towards heat death. And this is even more the case for a closed universe that periodically expands and contracts. The likelihood that neutrinos possess rest-mass, together with a number of other special features of the history of the universe, would tend to favor its being closed.

The structures of the universe—its galaxies, stars, planets, and, on earth, life—ultimately result from the export of entropy into the cosmic microwave background radiation, at the low temperature of 3°K.

Chapter 6

Open systems

Only in an open system
Can events occur,
Can time be without end,
Can discoveries be made.

The production and flow of entropy

So far we have considered, almost exclusively, isolated systems, that is, systems insulated from the external world by an impenetrable cocoon. All of our laws—including the first and second laws of thermodynamics—relate to isolated systems. If a deviation from the state of equilibrium should take place for some reason, that is, if the system should fluctuate to a state of non-maximal entropy, then, after a certain time, the system will revert to a state of equilibrium with entropy at a maximum.

But what if the system is open, that is, if it interchanges matter and/or energy with the surrounding world? Examples are not hard to find: each of us—every living organism, in fact—is such a system.

Clearly, in an open system not only does the second law fail, but also the first (the law of conservation of energy) and even the law of conservation of mass (Lomonosov's law) breaks down. A person eats, and thereby gains weight. One can gain or lose weight, and it behooves one (this is of great importance) to make an effort to maintain one's weight at an appropriate (not very high) level.

What does a change in the entropy of an open system involve?

First, the system's entropy can change—by an amount d_iS, say—as a result of various processes taking place internally. Second, entropy can flow into or out of the system. This might occur through heat exchange with the environment, or through interchange of matter. After all, entropy is a function of the state of matter, so that it will enter or leave the system together with the matter entering or leaving. Let d_eS denote the change in entropy resulting from the influx or efflux of matter. Thus the total change in entropy of an open system is given by

$$dS = d_iS + d_eS. \tag{6.1}$$

The sign of d_iS, the entropy change produced from within the system, is always positive, since if one imagines the system insulated by means of an adiabatic membrane, then d_eS will be zero, but the change d_iS will remain, and by the second law we must then have

$$dS = d_iS \geq 0. \tag{6.2}$$

Of course, we shall have equality here precisely if the (insulated) system has achieved equilibrium.

On the other hand, the sign of the change d_eS may be positive or negative, depending on circumstances, that is, on whether the influx of entropy is greater or less than the efflux. The net change dS in entropy of an open system may be positive, negative, or zero, depending on the contributions d_iS and d_eS to dS in (6.1):

$$\begin{aligned}
&d_eS > 0, && dS > 0;\\
&d_eS < 0, \quad \text{but} && |d_eS| < d_iS,\ dS > 0;\\
&d_eS < 0, \quad \text{and} && |d_eS| > d_iS,\ dS < 0;\\
&d_eS < 0, \quad \text{and} && |d_eS| = d_iS,\ dS = 0.
\end{aligned}$$

The last two of these four cases are the most interesting. They represent respectively the situations where an open system's entropy decreases because the efflux of entropy exceeds its creation internally, and where the efflux of entropy and internally created entropy exactly balance out.

In the situation where there is no change in the internal entropy, and only heat exchange with the environment occurs, then

$$dS = d_eS = \frac{dQ}{T}. \tag{6.3}$$

This is just the formula (2.4). Here the meaning of the expression "flow of entropy" is especially simple: it is essentially just the flow of heat. On the other hand, if entropy *is* produced internally, then the formula becomes

$$dS = d_iS + d_eS = d_iS + \frac{dQ}{T}, \tag{6.4}$$

which we also encountered earlier (see (2.30)) as an inequality holding for irreversible processes:

$$dS \geq \frac{dQ}{T}, \tag{6.5}$$

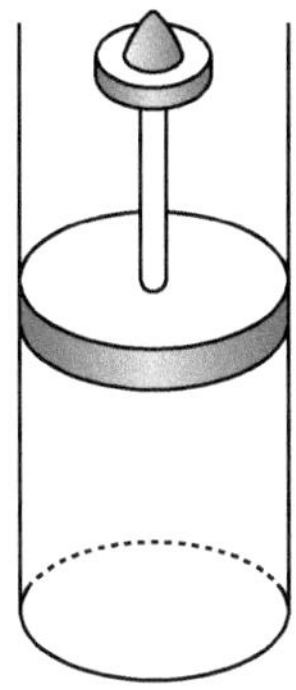

Figure 6.1: Model of quasi-equilibrial expansion of a gas.

with equality precisely if $d_iS = 0$.

As we saw in earlier chapters, entropy is always generated in systems within which physical, chemical, or biological processes take place. This production of entropy proceeds at a definite rate, a rate that is never negative, and becomes zero, that is, satisfies

$$\frac{d_iS}{dt} = 0, \tag{6.6}$$

only under conditions of internal equilibrium. Denoting by σ the amount of entropy produced internally per unit time per unit volume of an open system, we have

$$\frac{d_iS}{dt} = \int \sigma dV \geq 0. \tag{6.7}$$

The quantity σ is called the *(specific) dissipation function* of the system.

The formulae (6.6) and (6.7) differ fundamentally from those of thermodynamics and statistical mechanics which we derived and worked with earlier: formulae (6.6) and (6.7) involve the time! We are now concerned with the rate of generation of entropy with respect to time.

This signifies a transition from thermostatics, that is, classical thermodynamics, to dynamics and kinetics. Thermostatics concerns processes in equilibrium, or rather processes that take place sufficiently slowly for equilibrium to be attained at each stage. For example, we may model the equilibrial expansion of a gas in the following way: a piston moving in a cylinder supports a load consisting of a pile of sand (Figure 6.1), from which we remove one grain at a time.

Formulae (6.6) and (6.7) relate to non-equilibrial processes, and in this they represent a better approximation of reality. After all, in nature there are no equilibrial processes; such processes are mere idealizations.

The task of physics is to determine the factors on which the rate of generation of entropy and the associated specific dissipation function depend, and the nature of this dependence.

The dissipation function

As already noted, entropy is generated in all physical, chemical, and biological processes. Suppose a system—for the moment it is immaterial whether open or closed—contains two bodies at different temperatures in contact with one another. It is clear that in this system heat will flow from the hotter to the cooler body immediately upon contact, and that the flow will continue until the temperatures are equalized. In this situation entropy is generated as a result of heat conduction. Thus, to repeat, the driving force in the production of entropy is temperature difference. In this connection it is appropriate to mention the following very general physical principle:

Only differences give rise to effects, and these in turn yield new differences.

Returning to our theme, we observe that temperature differences give rise to a nonzero flow of thermal energy. In processes involving the generation of entropy, some sort of "flow" is always present—in the sense of a change in some physical quantity—together with a "force" causing the flow. In the present case, the flow is that of energy, namely dE/dt (where E is the thermal energy per unit volume), and the "force" is the temperature difference. However since the dependence of entropy on temperature is one of inverse proportionality, it is rather the difference of the *reciprocals* of the temperatures that is of significance for entropy:

$$\frac{1}{T_2} - \frac{1}{T_1} = \frac{T_1 - T_2}{T_1 T_2}, \quad T_1 > T_2.$$

Thus in the case of heat conduction, it is not difficult to see that the dissipation function has the form

$$\sigma = \frac{dE}{dt}\left(\frac{1}{T_2} - \frac{1}{T_1}\right). \tag{6.8}$$

Note that since σ is the quantity of entropy produced per unit volume per unit time and E is the energy per unit volume, the dimensions of the two sides of equation (6.8) coincide.

Observe also that one always has $\sigma > 0$. For, the flow of thermal energy dE/dt between the bodies at temperatures T_1 and T_2 is positive if $T_2 < T_1$, and then $1/T_2 - 1/T_1$ is also positive, while if $T_1 < T_2$, then both dE/dt and $1/T_2 - 1/T_1$ are negative. It turns out that the expression given by (6.8) for the dissipation function in this particular case as the product of the "flow" by the "force", reflects the general case.

Suppose, for instance, we are considering not heat conduction but electrical conduction. The "flow" would then be a flow of charge, that is, of electrons, and the "driving force" would be the electrical driving force, that is, a potential difference. Once again the effect—the electric current—has a difference as its cause. In this situation, the dissipation function turns out to be given by

$$\sigma = I\frac{\varphi_1 - \varphi_2}{T}, \tag{6.9}$$

where $I = de/dt$ is the current, which is just the rate of flow of charge per unit time, and $\varphi_1 - \varphi_2$ is the difference in potentials. The reader may recall that the quantity $I(\varphi_1 - \varphi_2)$ is *electric power*, representing the rate of transformation of energy, and usually measured in watts.[1] Hence for the dimensions to be those of the dissipation function (or *specific internal entropy*) σ, we need to divide the electric power by the temperature T, as in (6.9). Thus entropy is generated in electrical processes according to the formula (6.9).

Entropy is also generated by any chemical reactions taking place in a system. As discussed earlier (see the third section of Chapter 3), a chemical reaction proceeds from a state of higher (Gibbs) free energy to one of lower free energy. Gibbs defined the *chemical potential* of a combination of reagents as the rate of change of free energy per unit volume per unit time. Differences in chemical potential measure the propensity of a reaction to occur. Such a difference is called the *affinity* of the reaction, denoted by $\mathcal{A}$. In this, chemical, context, affinity plays the role of the driving "force", while the "flow" is represented by the rate of the reaction, that is, the number of moles of the reagents undergoing chemical transformation per unit time. The dissipation function then has the usual form:

$$\sigma = \frac{v\mathcal{A}}{T}. \tag{6.10}$$

Does the right-hand side here have the correct dimensions? Well, the rate of a reaction is measured in moles per unit time, while its affinity—the difference in chemical potentials—is measured in units of energy per mole per unit volume. Hence the dimensions of the right-hand side of (6.10) are those of

$$\frac{\text{moles}}{\text{time}} \cdot \frac{\text{energy}}{\text{mole·volume·temperature}} \equiv \frac{\text{energy}}{\text{time·volume·temperature}},$$

which are indeed those of the dissipation function.

The above-described processes are all irreversible. In general, in a non-equilibrial system several processes of various kinds may be going on at once.

In all cases, the "forces" and "flows" become zero at equilibrium; the entropy, having reached a maximum, stops increasing. In the first process above, we have $T_1 = T_2$, and the flow of thermal energy ceases, $dE/dt = 0$; in the second, $\varphi_1 = \varphi_2$, and the electric current ceases flowing, $I = 0$; and in the third, the affinity becomes zero, $\mathcal{A} = 0$, and the reaction stops, $v = 0$. For small perturbations from equilibrium, one may assume in each case that the "flow" is directly proportional to the "force". Thus in the case of heat conduction, the flow of thermal energy, dE/dt, is proportional to the temperature difference $T_1 - T_2$. In the case of electrical conduction, the electrical current is proportional to the potential difference; this is just Ohm's law:

$$I = g(\varphi_1 - \varphi_2), \tag{6.11}$$

[1] If potential difference is measured in volts and current in amps. *Trans.*

where g is the *electrical conductivity*, the reciprocal of the resistance. In a chemical reaction, the rate of the reaction is proportional to the affinity.

If a number of irreversible processes are taking place at the same time in an open, nonequilibrial system, then these processes turn out to be connected to one another, in the sense that the "flow" peculiar to any one of them may depend on the "forces" associated with the other processes.

Suppose for simplicity that we have such a system where just two irreversible processes are proceeding (for example, heat conduction and diffusion). Let J_1 and J_2 denote the respective "flows" and X_1 and X_2 the "forces". The dissipation function of the system is then given by

$$\sigma = J_1X_1 + J_2X_2 \geq 0. \tag{6.12}$$

It can be shown that, provided the system is close to equilibrium, the "flows" depend on the "forces" linearly:

$$\begin{aligned} J_1 &= L_{11}X_1 + L_{12}X_2, \\ J_2 &= L_{21}X_1 + L_{22}X_2, \end{aligned} \tag{6.13}$$

where the coefficients L_{11}, L_{12}, L_{21}, and L_{22} are constants of this particular system. It turns out that the coefficients of proportionality—"phenomenological coefficients"—are symmetric with respect to the indices 1 and 2:

$$L_{12} = L_{21}, \tag{6.14}$$

that is, that part of the flow J_1 attributable to the force X_2 is equal to the part of the flow J_2 attributable to X_1. The statement that this is so is called "Onsager's theorem".[2]

Substituting from (6.13) in (6.12), we obtain

$$\sigma = L_{11}X_1^2 + 2L_{12}X_1X_2 + L_{22}X_2^2 \geq 0. \tag{6.15}$$

Thus we have a positive definite quadratic form in X_1 and X_2. It follows that the coefficients L_{11} and L_{22} must be positive, since setting $X_2 = 0$ yields $\sigma = L_{11}X_1^2 > 0$ if $X_1 > 0$, whence $L_{11} > 0$, and similarly for L_{22}. It is easy to show (and well known) that the quadratic form in (6.15) is positive definite if and only if

$$L_{12}^2 < L_{11}L_{22}, \tag{6.16}$$

so this is the only limitation on the value of L_{12}.

(Note that, although of course the number n of processes taking place in a system may be arbitrarily large, the case $n = 2$ that we are presently considering involves no real loss of generality; the general case is analogous.)

---

[2] Lars Onsager (1903-1976), Norwegian/American physicist and chemist, 1968 Nobel laureate in chemistry.

The form of the expression for σ in (6.12) is highly significant, in that, although the sum σ of the terms J_1X_1 and J_2X_2 cannot be negative, one or other of the terms may be. For instance, if $J_1X_1 > 0$ but $J_2X_2 < 0$, then the condition $\sigma > 0$ is satisfied only if $J_1X_1 > |J_2X_2|$. This has the following important consequence. The negativity of J_2X_2 means that, taken in isolation, the corresponding process (No. 2) is impossible, since it would involve a decrease in entropy. Yet the fact that the other process (No. 1), for which $J_1X_1 > 0$, is taking place at the same time and producing surplus entropy sufficient to more than make up for that decrease, allows process No. 2 to proceed after all in the open system.

By way of example, we consider the surprising phenomenon of thermodiffusion. Suppose we have a vessel filled with a homogeneous mixture of two gases. If the temperature of the vessel is uniform, then the mixture will be in equilibrium, with its entropy at a maximum. However, if opposite walls of the vessel are at different temperatures, then there will occur a partial separation of the gases: one will tend to accumulate near the warmer wall and the other near the cooler. The loss of entropy occasioned by this separation is more than compensated by the gain in entropy due to the heat flow. The flow (in opposite directions) of the gaseous matter and the flow of the thermal energy are interconnected.

In this section we have discussed certain aspects of the so-called "linear thermodynamics of open systems", valid close to equilibrium. We have become acquainted with special features of such systems, in particular, with the possibility of processes involving a loss in entropy actually occurring as a consequence of their being interconnected with other "entropically advantageous" processes.

In the next section, we shall discuss other, equally important and interesting, peculiarities of open systems.

The area of non-equilibrial thermodynamics was developed in the work of L. Osager, T. de Donder,[3] and I. Prigogine.[4] However, the basic ideas of non-equilibrial linear thermodynamics had been described earlier in the work of L. I. Mandel′shtam and M.A. Leontovich entitled "Towards a theory of the absorption of sound in liquids", published in 1937.

An astronaut lives on negative entropy

This is all very interesting, important, and useful, but how can one apply the basic laws of thermodynamics to open systems? After all, those laws hold only for isolated systems.

The answer is rather simple: Along with the open system one is studying, one must consider its immediate environment—including all sources of matter and energy impinging on the system—, and imagine the resulting enlarged system to be separated from the surrounding world by means of an adiabatic, impermeable

[3]Théophile de Donder (1872–1957), Belgian mathematician and physicist.

[4]Ilya Prigogine (1917–2003), Russian-born, Belgian chemist. 1977 Nobel laureate in chemistry for his work in non-equilibrial thermodynamics.

shell. Then we can apply the laws of thermodynamics to this enlarged system in order to derive important results.

A good model of this is afforded by an astronaut in the cabin of a spaceship. Here, indeed, no effort has been spared to isolate the interior of the space ship from the surrounding space. The astronaut is secure in the cabin, with a supply of food, water, and air. Let's calculate the balance of entropy.

Taken alone, the astronaut represents an open system. Any infinitesimal change $dS^{(a)}$ in his or her entropy satisfies

$$dS^{(a)} = d_i S^{(a)} + d_e S,$$

where $d_e S$ is the entropic contribution from the surrounding medium (the interior of the space ship) via any interchange of thermal energy or matter. Hence the change in the entropy of the medium surrounding the astronaut is

$$dS^{(m)} = -d_e S.$$

The total change in entropy of the astronaut together with the interior of the space ship is

$$dS = dS^{(a)} + dS^{(m)} = d_i S^{(a)} > 0, \tag{6.17}$$

where, once again, the inequality obtains by virtue of the second law. Thus the entropy of the combined system of astronaut and space ship increases by an amount equal to the entropy generated by the astronaut's organism.

Now an astronaut must be a healthy young person, so that his or her state will remain the same for the whole of the flight. However though stationary, this state is not one of equilibrium. This means that the entropy (as well as the mass and energy) of the astronaut remains unchanged, whence

$$dS^{(a)} = d_i S^{(a)} + d_e S = 0. \tag{6.18}$$

Since by (6.17), $d_i S^{(a)} > 0$, it follows that $d_e S$ must be negative. We conclude that in a stationary open system the entropy generated within the system must be exactly balanced by the amount that flows out of the system. In fact, it can be shown that the entropy of the substances excreted from a living organism actually *exceeds* the entropy of the substances it consumes. In 1944, E. Schrödinger, one of the founding fathers of quantum mechanics, wrote a small book entitled *What is life?* (We mentioned this book earlier, at the end of Chapter 3). The appearance of this book was an event of considerable importance, since the ideas Schrödinger expressed in it played a significant role in the development of modern biology. In particular, the basic aspects of the thermodynamics of living (and therefore open) systems are discussed in the book.

What does an organism live on? We are used to talking of the number of calories we consume with food. Does this mean that we feed on calories, constantly adding energy to that already present in the organism? Of course not! If an organism is in a stationary state, then the amount of energy present in it is constant. To quote Schrödinger:

"What then is that precious something contained in our food that keeps us from death? That is easily answered. Every process, event, happening—call it what you will; in a word, everything that is going on in Nature means an increase of the entropy of the part of the world where it is going on. Thus a living organism continually increases its entropy—or, as you may say, produces positive entropy—and thus tends to approach the dangerous state of maximum entropy, which is death. It can only keep aloof from it, that is, alive, by continually drawing from its environment negative entropy—which is something very positive as we shall immediately see. What an organism feeds upon is negative entropy. Or, to put it less paradoxically, the essential thing in metabolism is that the organism succeeds in freeing itself from all the entropy it cannot help producing while alive."

Consumption of negative entropy means excretion of more entropy than enters the organism, signifying in turn the maintenance of a stationary state through an efflux of entropy.

Note that the views of Emden concerning heating a room (see the final two sections of Chapter 3) are fully analogous to those of Schrödinger quoted above.

A stationary state is possible only in an open system; such a state might be termed a "flowing equilibrium". An open system in a stationary state has a number of special features.

Earlier (in the third section of Chapter 2) we modeled a chemical reaction by the flow of a liquid from one container to another. We shall now represent an open chemical system by a model of a similar kind (Figure 6.2). This model is flowing (or "steady state") since the liquid is continuously replenished from below in the upper vessel and flows continuously downwards and out of the lower vessel.[5] As before, the role of the catalyst is played by the pipe connecting the vessels, regulated by means of a tap. If the system were, as earlier, isolated,[6] then the final state of the system, represented by the level of liquid in the lower vessel, would be independent of the extent to which the tap is open—the catalyst affects only the rate of the reaction, not the final outcome. By contrast, in the present open system the extent to which the taps are open determines not only the rate but also the level of liquid in the two containers, that is, the steady state of the process.

It was shown by Prigogine that if an open system is in a stationary state close to an equilibrium state, then the dissipation function is at its minimum value. In other words, the amount of entropy produced in a near-equilibrial stationary state is less than that produced in other states of the system. Therefore, as the system approaches a stationary state, its dissipation function decreases:

$$\frac{d\sigma}{dt} < 0,$$

[5]And is then pumped up to the higher vessel. *Trans.*

[6]With the liquid in the upper vessel not being replenished, and liquid not flowing out of the lower vessel. *Trans.*

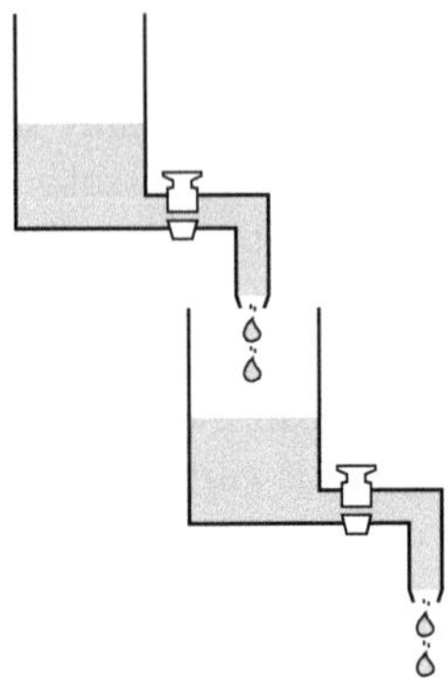

Figure 6.2: Model of a chemical reaction in an open system.

Figure 6.3: Dependence of the dissipation function on time: approach to a stationary state.

and, finally, takes on a minimum value, where

$$\frac{d\sigma}{dt} = 0,$$

as shown in Figure 6.3. Once reached, the stationary state is stable; a linear system will not leave such a state spontaneously. The steady-state system shown in Figure 6.2 behaves in this way: For given degrees of openness of the taps, the levels of liquid in the two vessels will achieve stability, provided the liquid continues to flow for long enough, and then for as long as it continues to flow.

The state of the biosphere as a whole can be regarded as stationary. The destabilizing effect of human activity has not yet significantly caused any change in the temperature or composition of the atmosphere.[7]

The realization of a stationary state in an open system requires two time scales, that is, the presence of a fast process and a slow one. We explain what we mean with the following simple example: Suppose we have two bodies at temperatures T_1 and T_2, that are connected by a fine copper wire, a good conductor of heat. Heat will flow along the wire until the temperatures of the bodies become equalized. The fast time scale is represented by the copper wire, which quickly enters a stationary state during which the rate of flow of heat through a cross-section of the wire is constant. The second, slow, time scale is represented by the process of equalization of the temperatures.

Of course, if we maintain the temperatures of the two bodies at T_1 and T_2 (by means of a heater and cooler, say), then equalization of the temperatures will not occur.

The stationary state of the biosphere provides another such example. This state arose and persists as a result, ultimately, of the stream of radiation from the sun. This state was established relatively rapidly[8] as a consequence of the appearance of green plant growth; photosynthesis in these organisms brought about the oxygenation of the atmosphere.[9] On the other hand, the "slow" time scale is here represented by the much slower cosmic process of the burning out of the sun.

We shall now imagine the crew of our space ship to consist of a baby and an old man. Neither of these two organisms is in a stationary state. As the baby grows, so do its mass and energy, and even the degree of its organic order. Thus

[7] If this could be said in 1986 (the date of publication of the original of this book)—which is doubtful—, it certainly cannot be seriously maintained now. Global warming due to the "greenhouse effect" of vast accumulations of carbon dioxide in the atmosphere, and the countervailing dimming of the sunlight falling on most places on earth due to atmospheric pollution, are concerns on everybody's mind in 2008. *Trans.*

[8] That is, over a period of more than a billion years. *Trans.*

[9] The process of oxygenation of the atmosphere is now thought to be much more complicated, involving so-called monocellular "blue-green algae" dating from about 2.5 billion years ago—when the atmospheric oxygen had reached 10% of its present level—and other even older single-celled organisms, to about 600 million years ago when the first multicellular organisms appeared, and when the atmospheric oxygen had reached about 90% of its present level. See *Encyclopedia Britannica*. *Trans.*

the baby's entropy is decreasing:

$$dS^{(baby)} = d_i S^{(baby)} + d_e S < 0, \tag{6.19}$$

that is, owing to a more powerful metabolism, the efflux of entropy exceeds the amount produced, and the baby absorbs more "negative entropy".

On the other hand, aging is accompanied by an increase in entropy that is not balanced by its efflux into the surroundings:

$$dS^{(old\ man)} = d_i S^{(old\ man)} + d_e S > 0. \tag{6.20}$$

As an elderly physicist once wrote:

Entropy's consuming me
Bit by bit and altogether.
My last dull years are measured off
Gram by gram and meter by meter.

In the case of a living organism, entropy attains its maximum in the equilibrium state—otherwise known as death.

Why do cells divide?

Elementary considerations of the thermodynamics of open systems allow us to understand why a living cell divides.

A cell is an open system, so the balance of any change in entropy is given by (6.1), that is:

$$\Delta S = \Delta_i S + \Delta_e S.$$

For the sake of simplicity, we shall assume that a cell is a sphere of radius r. The amount of entropy $\Delta_i S$ produced per unit time inside such a sphere is proportional to its volume $\frac{4}{3}\pi r^3$, and the efflux $\Delta_e S$ of entropy is proportional to its surface area $4\pi r^2$. Hence

$$\Delta S = A \cdot \frac{4}{3}\pi r^3 - B \cdot 4\pi r^2,$$

where A and B are constants of appropriate dimensions. As the cell grows, r increases, and when $r = 3B/A$, the cell enters a stationary state: $\Delta S = 0$. For smaller values of r we have $\Delta S < 0$, that is, the efflux of entropy exceeds the amount generated within the cell, and the cell can grow. However, for $r > 3B/A$, we have $\Delta S > 0$, which means that substances with redundant entropy are accumulating in the cell and causing it to overheat. Hence when r reaches the value $3B/A$, the cell must either divide or perish. On division, while the total volume remains unchanged, the combined surface area of the two daughter cells is greater than that of the mother cell. This can be seen as follows. Denoting by r' the radius of each daughter cell, the equality of volumes before and after the division yields $r^3 = 2r'^3$, whence

$$r' = \frac{r}{\sqrt[3]{2}}.$$

The new entropy change per unit time is

$$\Delta S = A \cdot \frac{8}{3}\pi r'^3 - 2B \cdot 4\pi r'^2,$$

and when $r = 3B/A$, we have $r' = 3B/(A\sqrt[3]{2})$, whence

$$\Delta S = 36\pi \frac{B^3}{A^2}(1 - \sqrt[3]{2}) < 0.$$

As a result of division, the efflux of entropy exceeds the amount generated within the cell, by the factor

$$\frac{|\Delta_e S|}{\Delta_i S} = \sqrt[3]{2} \approx 1.26.$$

Although thermodynamics explains why a cell must divide, it tells us nothing about the actual mechanism of this extremely complicated process. Thermodynamics is a phenomenological science.

The growth of a living organism, which always consists of cells, is fundamentally different from the formation of a crystal from the liquid state or a solution. The cell-division and resulting growth of an organism are directly related to the efflux of entropy into the surrounding medium. Such processes are non-equilibrial. On the other hand, the growth of a crystal is an equilibrial process that takes place when the free energies of crystal and liquid have become equal, and thus amounts to an equilibrial phase transition. Biological development, while it resembles phase transition, is fundamentally non-equilibrial.

Far from equilibrium

The development of an embryo and, as noted above, the subsequent post-partum growth of the baby, involve an increase in order. The embryo starts as a single fertilized cell—an ovum—and its further development—ontogenesis—is accompanied by morphogenesis, that is, the formation of various specific structures—tissues and organs. Another example of structure-formation is afforded by the rise of galaxies and stars in the cosmos.

Such processes necessarily involve decreases in the entropy of the relevant open systems, that is, the export of entropy into the surrounding medium in accordance with the inequality (6.19):

$$d_e S < 0, \ |d_e S| > d_i S > 0. \tag{6.21}$$

These conditions can hold only in states far from equilibrium—since the term $d_i S$ dominates near equilibrium; the extreme case here would be represented by a stationary state. For example, our astronaut (see above) was young and healthy, and very far from equilibrium.

In order for structure-formation to take place in an open system, that is, for a radical increase in order to occur, the export of entropy must exceed a certain critical value.

For the export of entropy to exceed its internal production, a sort of "entropy pump" is needed, to pump entropy out of the open system. Such a "pump" can work either externally or internally. We shall now find the thermodynamic conditions for the effective working of such a pump.

An infinitesimal change in the (Helmholtz) free energy of an open system at fixed temperature and volume is given by (see equation (6.1) and the discussion preceding and following it)

$$dF = d_iF + d_eF = dE - TdS = d_iE + d_eE - Td_iS - Td_eS.$$

Now $d_iE = 0$ since the energy of the system can change only via an interchange of energy with the surrounding medium and not internally. Hence

$$d_eF = d_eE - Td_eS = dE - Td_eS,$$

and then conditions (6.21) imply that

$$d_eF > dE + Td_iS. \tag{6.22}$$

Thus in order for an export of entropy to occur, that is, an efflux from the system, an amount of free energy is needed greater than the change in internal energy[10] plus the contribution resulting from the production of entropy within the system.

If we are considering instead a process taking place at constant pressure rather than constant volume, then it is appropriate to use the Gibbs free energy in place of the Helmholtz free energy. The relevant inequality is then

$$d_eG > dH + Td_iS. \tag{6.23}$$

If the system is in a stationary state, then its internal energy does not change (and therefore neither does its enthalpy) :

$$dE = dH = 0,$$

and the above inequalities become

$$d_eF = d_eG > Td_iS > 0, \tag{6.24}$$

showing that one needs to supply free energy to the system in order to maintain the stationary state.

[10] That is, the internal energy of the system together with its active environment. *Trans.*

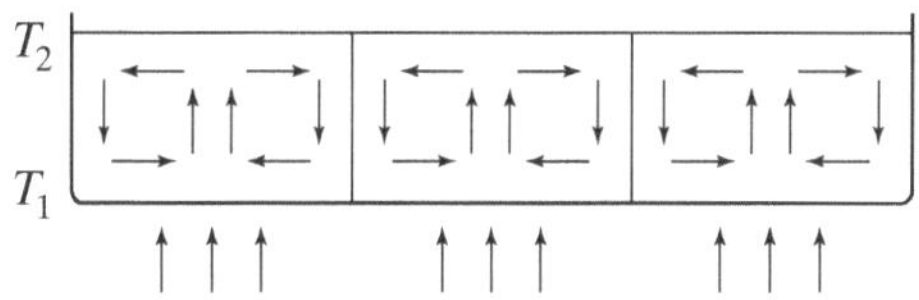

Figure 6.4: How Bénard convection arises.

"Bénard convection" [11] affords a beautiful example of structure-formation under relatively simple conditions.

We have a shallow vessel containing a viscous fluid such as silicone oil. We heat the vessel strongly from below, thereby causing a temperature difference $\Delta T = T_1 - T_2 > 0$, between the lower and upper surfaces of the liquid. As long as ΔT is small, the liquid remains untroubled, and heat is transferred from the bottom to the top by heat conduction. Then, at a certain critical temperature difference ΔT_{cr}, there is a sudden change in the behavior of the liquid: convection sets in (Figure 6.4) and the liquid separates itself into hexagonal cells (Figure 6.5). The result is very attractive and truly remarkable: as a result of heating the liquid there arises a dynamic, organized structure, resembling a crystalline one. It is created by the simultaneous cooperative motions of the molecules of the liquid.

Figure 6.6 shows the dependence of the rate of heat transfer dQ/dt on the temperature difference ΔT. At the critical temperature difference ΔT_{cr} there occurs a sudden change in the dependence, signalling the formation of a cellular structure.

Since this system absorbs heat from its surroundings, the rate of flow of entropy through the exposed surface of the liquid is given by the formula (see (6.3))

$$\frac{d_e S}{dt} = \frac{dQ}{dt}\left(\frac{1}{T_1} - \frac{1}{T_2}\right) = \frac{dQ}{dt}\frac{T_2 - T_1}{T_1 T_2} < 0. \tag{6.25}$$

Hence under these conditions the system exports entropy. Under stationary conditions this export or efflux of entropy exactly balances the amount of entropy generated within the liquid through internal friction and heat conduction.

A more detailed analysis shows that the surface tension of the liquid plays an essential role in Bénard's effect.

Prigogine called open systems that are structure-forming, self-organizing, and far from equilibrium *dissipative*. Such systems form spatially (see below), as well as temporally, stable structures as a certain parameter passes through a critical value. In the case of Bénard's effect, the parameter is the temperature difference.

An example of a completely different sort is afforded by lasers. For the sake of concreteness, we shall consider only solid state lasers, for example ruby lasers. A ruby laser consists of a cylindrical ruby rod whose ends are silvered. A pulsating xenon lamp is used for "optical pumping" of the laser: light from the xenon lamp

[11] Henri Bénard (1874–1939), French physicist.

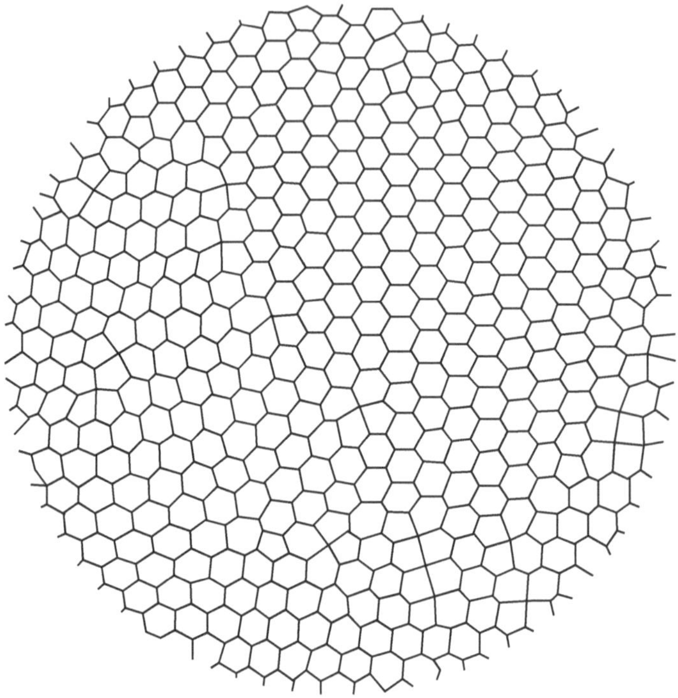

Figure 6.5: Bénard convection.

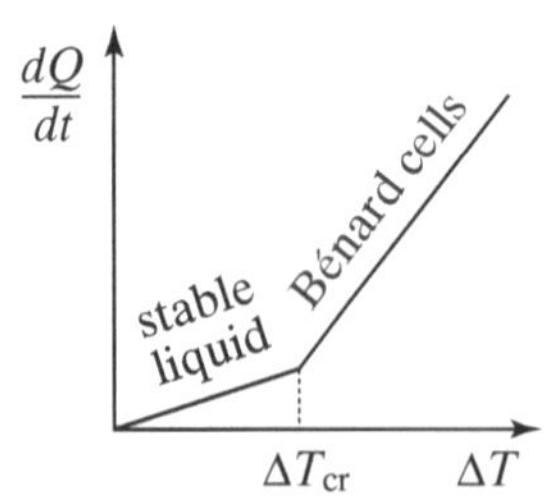

Figure 6.6: The dependence of the rate of heat transfer on the temperature difference.

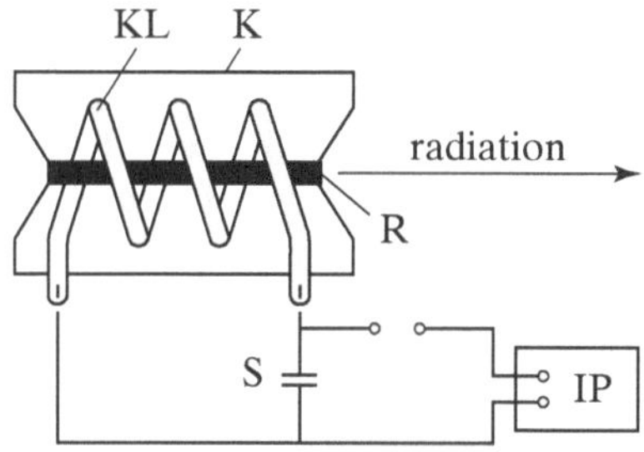

Figure 6.7: Diagram of a ruby laser: R is a synthetic ruby rod, K the cover, KL a xenon flash lamp for optical pumping, S a capacitor, and IP a source of constant voltage.

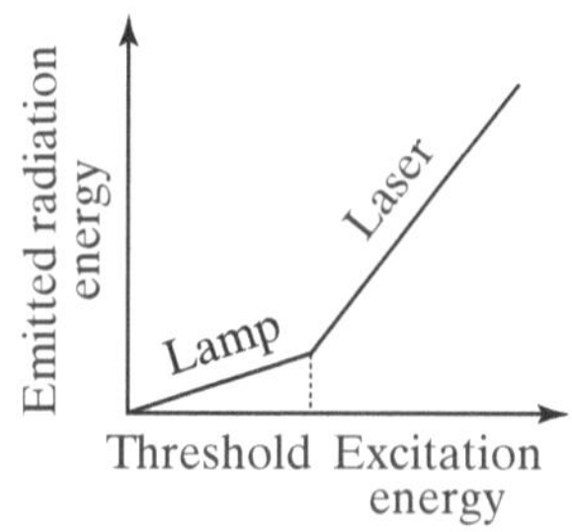

Figure 6.8: The dependence of the emitted radiant energy on the excitation energy of the laser.

is absorbed by the ruby rod mainly at wavelengths around 410 and 560 nanometers in two absorption bands. The optical pumping causes pulsating radiation of wavelength 694.3 nm to be emitted through the ends of the rod (Figure 6.7).

The xenon lamp optically excites the ions Cr^{3+} of trivalent chromium responsible for the absorption spectrum of rubies, that is, for their color, and these then emit pulses of light a few meters in length. Each pulse lasts about 10^{-8} seconds. The mirrors, that is, the silvered ends of the ruby rod, emit the radiation parallel to the rod's axis. At small amounts of optical pumping, the laser acts like a lamp, since the emissions from the individual excited ions do not cohere. However, at a certain critical value of the energy of discharge of the xenon lamp and a critical (threshold) value of the frequency of pulsation, the power of the laser emission increases abruptly. The irradiated ions now give off light coherently—in cooperation, as it were—, emitting waves in phase with one another. The lengths of the pulses grows to 10^8 or 10^9 meters, and the ruby laser goes over from a regime of ordinary radiation, as from a lamp, to one of laser radiation. Of course, this process is very far from equilibrium. The schematic graph of Figure 6.8 shows the transition from ordinary lamplike emission to laser emission. Observe the similarity to the graph in Figure 6.6 relating to Bénard's effect.

One could say that the most important things in the universe arose in the form of dissipative ordered structures far from equilibrium as a result of the export of entropy. In particular, the galaxies and stars originated in this way: gravitational energy was transformed into thermal energy, leading to a local efflux of entropy. The biosphere as a whole, and each living organism in particular, are highly ordered and far from equilibrium; thus life exists on earth as a consequence of the export of entropy.

In the most highly self-organizing systems, in the course of their evolution over time (relative to the same value of the mean energy) their entropy decreases—the amount of entropy exported comes to exceed the amount produced. The Soviet physicist Yu.L. Klimontovich[12] called this assertion the "S-theorem"; thus while Boltzmann's "H-theorem" (see the section entitled "The age of Darwin" in Chapter 5) relates to equilibrium systems, the S-theorem has to do with dissipative ones.

The basic features peculiar to dissipative structures are as follows. First, they occur in open systems far from equilibrium and arise as a result of an increase in fluctuations—that is, small deviations from the most likely state—rising to a macroscopic level. In this way order is created out of disorder, out of chaos. This kind of order is fundamentally different from the ordinary crystalline order that arises under equilibrium conditions. The difference consists precisely in the disequilibrium of a dissipative system, maintained by a forced export of entropy. All the same, the appearance of spatial or temporal order in a dissipative system is analogous to a phase transition. Phase transitions under conditions of equilibrium, such as, for instance, crystallization, are due solely to the interactions of a multitude of particles, whereby the particles effect the change of state coherently, as it were cooperatively. On the other hand, the transition of a dissipative system to an ordered state from a prior unstable disordered state takes place when some parameter reaches a critical value. It is precisely in such situations that small fluctuations grow to a macroscopic level.

Phase transitions constitute an important and far from simple area of physics. They are "cooperative" phenomena. The failure to realize this has led in the past to very serious errors. For example, it was once claimed that the dependence on temperature of the ratio of the number of molecules in the solid (crystalline) state to the number still in the liquid state can be derived from the Boltzmann distribution (4.20). Let N_{crys} denote the number of crystallized molecules in a given quantity of the substance in question at temperature T, and N_{liq} the number in the liquid state. If we blindly apply the Boltzmann distribution (4.20), then we obtain

$$\frac{N_{crys}}{N} = \frac{\exp(-E_{crys}/kT)}{\exp(-E_{crys}/kT) + \exp(-E_{liq}/kT)},$$

[12]Yuri L′vovich Klimontovich (1924(?)–2004(?)), Soviet/Russian physicist.

where N is the total number of molecules. Hence

$$\frac{N_{liq}}{N_{crys}} = \frac{N - N_{crys}}{N_{crys}} = \exp\left(-\frac{E_{liq} - E_{crys}}{kT}\right).$$

However, this makes no sense, since according to this formula the substance will freeze completely only in the limit as $T \to 0°\mathrm{K}$, while as the temperature approaches ∞ the two sorts of molecules become equal in number:

$$N_{liq} = N_{crys}.$$

The mistake here arises from failing to take into account the interaction of the molecules and their coordinated behavior. The statistical sum has to be modified to accomodate this feature.

The transition to a dissipative structure in an open system is a *non-equilibrial* phase transition. Haken[13] called the area of physics concerned with coordinated phase transitions—both equilibrial and non-equilibrial, but principally the latter—"synergetics".

The Belousov–Zhabotinskiĭ reaction

It can be shown that if a system deviates just a little from equilibrium, then its return to the equilibrium state will proceed smoothly without oscillations according to the following exponential law: If Δa measures the deviation of some physical parameter from its equilibrium value a_e, then

$$\Delta a = a(t) - a_e = (a(0) - a_e)e^{-t/\tau},$$

where τ is a constant called the *relaxation time* of the system. Thus as $t \to \infty$, $\Delta a \to 0$, that is, the perturbed value $a(t)$ of the parameter tends to the equilibrium value.

If the system is in a stationary state close to equilibrium, and the system deviates slightly from that stationary state, it will return to that state in accordance with the same law.

The relaxation time τ determines the rate at which the system returns to equilibrium or to the stationary state. At $t = \tau$ we have

$$a(\tau) - a_e = \frac{a(0) - a_e}{e},$$

that is, the initial deviation has by time τ decreased e (≈ 2.78) times.

On the other hand, in certain systems far from equilibrium, where, as we have seen, dissipative spatial and temporal structures—inequilibrial order—can arise, this order may consist of oscillations or waves. This is especially striking in certain dissipative chemical systems.

[13] Hermann Haken (born 1927), German theoretical physicist.

Imagine a lecturer demonstrating a chemical experiment. He fills a beaker with a blue liquid, and using a pipette adds a few drops of a colorless liquid. The solution in the beaker turns pink.

Well, there's nothing so remarkable about that. In chemistry there are many much more surprising transformations!

But wait! What's happening? After about a minute the liquid in the beaker turns blue again, then pink again, then blue again, and so on. The liquid's color changes periodically, like a sort of chemical clock.

This remarkable phenomenon—a periodic chemical reaction in a homogeneous solution—was discovered by B.P. Belousov[14] in 1951. In 1910 Lotka[15] showed in important theoretical work that in a chemical system far from equilibrium, oscillations in the degrees of concentration of the reagents are possible. Then in 1921 Bray[16] observed for the first time a periodic chemical reaction in a solution of hydrogen peroxide H_2O_2, iodic acid HIO_3, and sulphuric acid H_2SO_4. He observed a periodic oscillation in the concentration of iodine in the solution as successive oxidations of iodine to iodate and then reductions back to iodine took place:

$$\begin{aligned} 5H_2O_2 + 2HIO_3 &\rightarrow 5O_2 + I_2 + 6H_2O, \\ 5H_2O_2 + I_2 &\rightarrow 2HIO_3 + 4H_2O. \end{aligned}$$

This reaction was rather complex, and for some time it could not be ruled out that it might be heterogenous,[17] taking place on bubbles of the iodate catalyst. Belousov discovered *his* reaction, subsequently to make him famous, sometime in the 1950s, and independently of the work of these scientists, but at the time was unable to publish it except in an obscure journal. It was intensively investigated further, starting in 1961, by A.M. Zhabotinskiĭ,[18] who was able to simplify the reaction, whence it became known as the "Belousov-Zhabotinskiĭ reaction". In this reaction the color change (between yellow and clear) is caused by a change in the charge on a metallic ion. We give a simplified description of Belousov's reaction, reducing it to just two stages. At the first stage, trivalent cerium is oxidized by bromic acid:

$$Ce^{3+} \xrightarrow{HBrO_3} Ce^{4+},$$

and the second consists in the reduction of the four-valent cerium by the organic compound malonic acid:

$$Ce^{4+} \xrightarrow{\textit{malonic acid}} Ce^{3+}.$$

---

[14]Boris Pavlovich Belousov (1893–1970), Soviet chemist and biophysicist.

[15]Alfred James Lotka (L′viv 1880–1949), American mathematician, statistician, and physical chemist.

[16]William Crowell Bray (1979–1946), Canadian physical chemist.

[17]That is, caused by reagents' being in two or more different phases. *Trans.*

[18]Anatol M. Zhabotinsky currently holds a position at a university in the U. S.

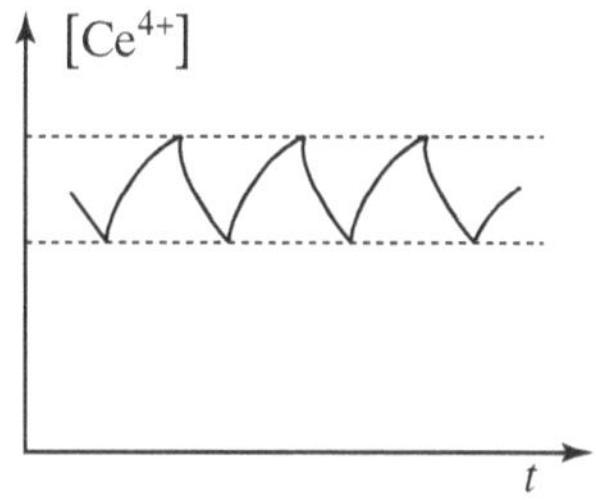

Figure 6.9: Variation of the concentration of Ce^{4+} in the Belousov-Zhabotinskiĭ reaction.

This periodic process comes to an end after a large number of oscillations as a result of irreversible exhaustion of the supply of the bromate anion BrO_3^-. The final products of the reaction are CO_2, H_2O, and bromium derivatives of malonic acid. No *perpetuum mobile* is achieved! Figure 6.9 shows the oscillations in the concentration of the four-valent cerium ion.

Zhabotinskiĭ and A.N. Zaikin later discovered and investigated other reactions of this type. But that is not all: By having periodic reactions take place in narrow tubes (one-dimensional systems) and in thin layers of solution (two-dimensional systems) in the absence of convection, Zhabotinskiĭ and Zaikin were also able to produce wavelike chemical processes. Figure 6.10 shows the evolution of such a wave in two dimensions. First, a germinating or initiating center—a germ or eye—appears as the result of a local fluctuation in the concentration, and from this center waves of color emanate according to variations in concentration.

Such phenomena of spatio-temporal ordering represent auto-oscillatory and auto-wave producing processes. Such processes occur in open nonlinear systems that are far from equilibrium, as a result of forces depending on the state of motion of the system itself, and furthermore the amplitude of the oscillations is determined by properties of the system, and not by any initial conditions. Auto-oscillatory and auto-wave processes in chemistry (and also in biology—see Chapter 8) are sustained by an efflux of entropy from the relevant system.

If a break occurs in the wave front of a chemical wave, then a spiral wave called a "reverberator" may result. Reverberators form, in particular, when two-dimensional waves propagate near an opening. Figure 6.11 is a photograph of chemical reverberators. Figure 6.12 is a photograph of a certain species of lichen—by way of comparison.

Organisms as dissipative systems

Of course, the surprising similarity of the above photographs does not mean that the growth of a lichen is in all respects like the propagation of a spiral chemical

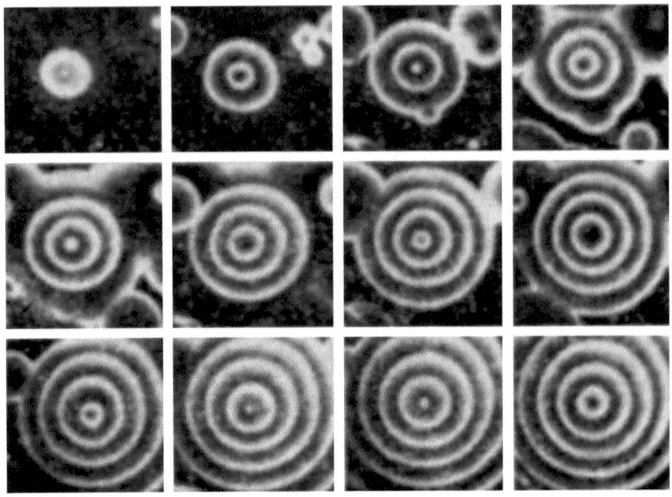

Figure 6.10: Initiating centers, and subsequent stages.

Figure 6.11: Chemical reverberators.

Figure 6.12: The lichen *parmelia centrifuga*.

wave. Nevertheless, at the heart of all biological phenomena we find the physics of open systems far from equilibrium. Although we shall be discussing this topic in detail in the final chapter of the book, it is appropriate at our present juncture to consider certain special structural and dynamical features of living organisms.

We have seen that in open systems far from equilibrium there can arise a specific structuring of the system as a result of intensification of fluctuations up to the macroscopic level, resulting from the efflux of entropy from the system.

In 1952, Turing[19] showed that the conjunction of an autocatalytic reaction with diffusion can cause spatial and temporal order to arise. A reaction is called *autocatalytic* if at least one of the products of the reaction is also a reagent, so that the quantity of that reagent increases with time. Probably the best known example is the replication of DNA (deoxyribonucleic acid) macromolecules that takes place in cell-division. The initial double helix of DNA catalyzes, via a process called "matrix synthesis", the formation of a copy of itself.

Turing's reaction-diffusion model is based on a reaction of the following type:

$$\mathrm{A} \to \mathrm{X}, \ \ 2\mathrm{X} + \mathrm{Y} \to 3\mathrm{X}, \ \ \mathrm{B} + \mathrm{X} \to \mathrm{D} + \mathrm{Y}, \ \ \mathrm{X} \to \mathrm{E}.$$

Here A and B are the initial reagents, X and Y the intermediate ones, and D and E the final products. The second stage of the reaction is the autocatalytic one: As a result of the action of the substance X, the substance Y is transformed into X, which thus catalyzes its own production. It follows easily that the overall reaction is, in sum,

$$\mathrm{A} + \mathrm{B} \to \mathrm{D} + \mathrm{E}.$$

Prigogine and his collaborators called such chemical systems "brusselators" for the simple reason that their early theoretical investigation was undertaken in Brussels.

The equations describing the kinetic behavior of brusselators and the diffusion of the reagents X and Y (brusselators in a distributed system) are nonlinear, and such a system is far from equilibrium. The concentrations of the substances X and Y undergo periodic oscillations, thus forming waves of varying concentration. At a certain threshold value of the concentration[20] an initial fluctuation from the stationary state is reinforced, ultimately bringing the system into a new stationary state corresponding to the new inhomogeneous distribution of the substances X and Y. Figure 6.13 shows one of the solutions of the kinetic equations of a brusselator: the distribution of the concentration [X] with respect to a spatial coordinate.

Turing's original article had the arresting title "The chemical basis of morphogenesis".[21] The term "morphogenesis" refers to the initiation and development of an organism's complex structure in the course of its embryonic growth, that is,

---

[19] Alan Mathison Turing (1912–1954), English mathematician, logician, cryptographer, and "the father of modern computer science".

[20] Of substance X? *Trans.*

[21] *Phil. Transactions of the Royal Soc. London. Series B, Biological sciences*, Vol. 237 (1952), pp. 37–72.

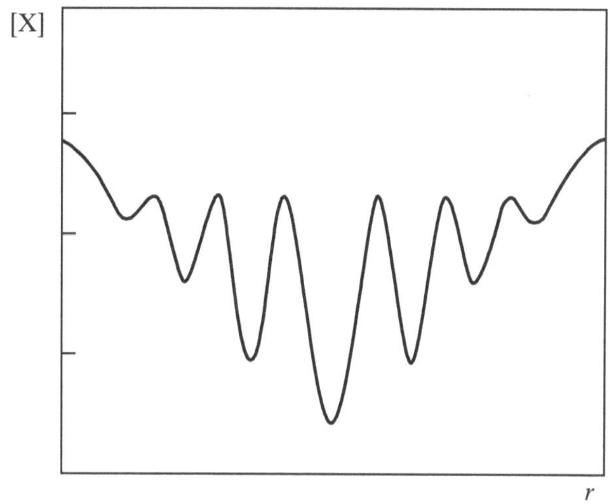

Figure 6.13: A localized stationary dissipative structure.

the differentiation of cells into tissues and organs. Turing was the first to establish the possibility that morphogenesis has a chemical basis. Of course, both Turing's theory and that of brusselators were based on chemical models, for all that very convincing. Now we know for a fact that morphogenesis in nature really is determined by molecular interactions, and that certain substances, namely "morphogenes", functioning at specific times and at specific places within the organism, are responsible for the formation of the various organic structures.

These interesting natural phenomena have much in common with auto-oscillatory and auto-wave processes, studied in especially great detail in the case of the Belousov-Zhabotinskiĭ reaction.

Occasionally one hears of claims to the effect that chemical auto-oscillatory processes and standing waves (such as the so-called "Liesegang rings" [22] observed in colloidal suspensions) can be explained only on the basis of quantum mechanics, and even that the Planck constant h can be determined from the periods of such oscillations. However this is a crude misapprehension: one might attempt with equal success to estimate Planck's constant from the zebra's or tiger's stripes. As we have seen, there is no trace of quantum mechanics in the theory underlying these periodic phenomena; they are macroscopic physical phenomena, not atomic or subatomic.

Many of the tissues of living organisms are *excitable*, meaning that excitation

[22] Any of a series of usually concentric bands of a precipitate (an insoluble substance formed from a solution) appearing in gels (coagulated colloid solutions). The bands strikingly resemble those occurring in many minerals, such as agate, and are believed to explain such mineral formations. The rings are named for their discoverer, the German chemist Raphael Eduard Liesegang (1869–1947). *Encyclopedia Britannica.*

—chemical or electrochemical—is transmitted across them from point to point, propagating like a wave. Muscular and nervous tissues have this property: the propagation of a stimulus along a nerve fiber and the synchronous oscillations of the whole heart muscle are phenomena with a chemical basis.

The Soviet physicist V.I. Krinskiĭ[23] has investigated one of the most dangerous of human pathologies, heart fibrillation.[24] In such fibrillation the heart has departed from its normal regime of regular contractions and entered one of chaotic oscillations, and without emergency medical aid the condition results in death. It turns out that this type of "cardiac arrhythmia" results from the multiplication of spiral waves of excitation, that is, of reverberators. In his works, not only does Krinskiĭ construct a theoretical model of the mechanism of fibrillations and report on his investigations of the phenomenon by means of delicate experiments on the hearts of suitable animals, but also gives practical advice as to what one should do to bring the fibrillations to an end in the event of an attack.

The three stages of thermodynamics

We have by now become thoroughly acquainted with the development of thermodynamics from its origins almost two hundred years ago. We end this chapter with an overview of the conceptual content of this development.

Prior to the birth of thermodynamics, science was dominated by Newtonian mechanics—a mechanics where time was reversible and the world did not evolve. At some point the Almighty set the universe's mechanism going, and thenceforth it has worked unchangingly like a wound-up clock. Living nature also appeared to be unchanging and unchangeable, remaining as it was at its initial creation. The founder of scientific biology, and the author of the classification of life forms, Carl Linnaeus,[25] considered that the biological life forms all about him were immutable, and created simultaneously at some time in the past.

The natural sciences as they were then did not involve time. Time was the prerogative of the humanities, above all history. There it was clear that time changes everything.

M.V. Lomonosov was an opponent of Newtonian physics, espousing Descartes' theory instead. Perhaps his rejection of Newtonian physics had something to do with the breadth of his interests, that is, with the fact that he was not only a physicist and chemist, but also a poet and historian; in poetry and history time flowed irreversibly. At the turn of the 18th century the first scientific-technological revolution occurred—or rather technological-scientific revolution, since the steam engine was invented independently of physics, and it all started from that. Sadi Carnot's thoughts on the steam engine represented the founding of thermodynamics. The

[23]Valentin I. Krinsky is now working at an institute in Nice, France.

[24]The lay term is "heart palpitations". *Trans.*

[25]Or Carl von Linné (1707–1778), Swedish botanist, physician, and zoologist. The father of modern taxonomy.

first and second laws of thermodynamics were discovered, and then somewhat later a third—Nernst's theorem. And entropy appeared on the scene—initially as the gray shadow of energy, the ruler of the universe.

Time first put in appearance in the second law of thermodynamics, formulated as the irreversible growth of entropy in spontaneous processes. Apart from this, however, thermodynamics remained thermostatics, the science of equilibrium and equilibrial processes.

Then Thomson predicted the heat death of the universe. It followed that the world cannot be standing still, it must be evolving; it moves inexorably towards its demise just like every living organism.

In this way the emphasis moved from technology to cosmology, representing a shift in focus from what exists to what develops—"from being to becoming", in the words of Prigogine. It was the dawn of the age of Darwin: ideas from biology (and the humanities) of the development and growth towards the most probable state of a physical system were incorporated into physics. The first stage of thermodynamics culminated in the creation of statistical physics in the works of Boltzmann and Gibbs. At this stage, entropy ceased to represent merely a measure of the depreciation of energy, and assumed its true role as a measure of the degree of disorder of a system, an objective characterization of the unavailability *in principle* of information about a system. The importance of entropy as one of the chief characteristics of arbitrary systems began to grow rapidly.

During the second stage in the development of thermodynamics, scientists turned to the study of open non-equilibrial systems close to equilibrium. This linear thermodynamics of open systems was created by Onsager, Prigogine, and others of their contemporaries. In this science, the dependence on time had become quantitative: as we saw, this non-equilibrial thermodynamics does not limit itself to the mere assertion that entropy increases in irreversible processes, but actually involves explicit computation of the rate of this increase, that is, of the derivative of the entropy content with respect to time—the dissipation function.

There are two fundamental features of linear thermodynamics (now definitely thermo*dynamics* and not thermostatics) that are nontrivial and essential. First, there now arises the possibility of an open system's existing in a stationary, but non-equilibrium, state, in which the production of entropy is balanced by its efflux from the system. Second, this thermodynamics allows for the conjunction of dynamical processes in a single open system, whereby a process that could not take place by itself (inasmuch as it involves a decrease in entropy), is realized through the free energy made available by other, entropically advantageous, processes.

Finally, the last 20 or 30 years[26] have witnessed a third stage in the evolution of thermodynamics, represented by the physics of non-equilibrial dissipative processes. We have seen in the present chapter that open systems far from equilibrium possess remarkable properties: they are capable of creating order from chaos by exporting entropy, that is, through its efflux out of the system. A living organism

[26]That is, starting in the 1960s.

feeds on negative entropy and not on positive energy. Thus at this stage, entropy has been promoted from the mere shadow of an omnipotent sovereign to a powerful entity determining the very existence of life on earth, and the evolution of the universe.

For the first time we are in a position to understand how it is that order, analogous to the crystalline variety, is possible in open systems far from equilibrium—that order can emerge out of chaos much like a phase change. A new branch of science has been born, the physics of dissipative systems (as Prigogine calls it) or synergetics (as Haken calls it). This relatively new area of physics holds out great promise.

The creation of this branch of science began the process of integration of the sciences characteristic of our time, superseding the former rigid specialization. Investigations of the Bénard effect, the laser, of periodic chemical processes, and heart fibrillation, are all pursued today from a unified scientific point of view. From the same unified stance, experts in synergetics investigate structure-formation in plasma[27] "on the earth, in the heavens, and on the sea". Yes, yes, from the periodicity sometimes evident in cloud formations to the northern lights (*aurora borealis*)—all such phenomena result from a single law of nature, namely the emergence of order from chaos on the analogy of a phase change. In the last analysis, cosmology itself is merely a part of the physics of dissipative systems.

At last, a century after the appearance of Darwin's *The origin of species*, physics has become united with biology in the task of comprehending the essence of irreversible processes. This will be discussed further in Chapter 8.

It might justifiably be claimed that the publication of *The origin of species* heralded the appearance of synergetics in science. Darwin showed how the orderly process of evolution—culminating, for us, in the present biosphere—arises out of the chaotic, disorderly mutability of natural living organisms. It is in this sense that one might call Darwin the founding father of synergetics.

Simultaneously with the rise of this nonlinear thermodynamics, a theory of information was being formulated, closely allied to thermodynamics. We have more than once spoken of entropy as a measure of the unavailability of information. But what exactly *is* "information"?

[27] In the general sense of a gas or suspension. *Trans.*

Chapter 7

Information

We understand, even though half-heard,
Of questions, answers every word.
But if those half-heard words are new,
They're irredundant through and through.

Information and probability

What *is* information?

The everyday meaning of the word is clear: It is what is communicated. We obtain it using every sense organ. We transmit it to others. Our personal life and the functioning of society are based on communication, on the receipt and transmission of information—and this applies not only to human beings, but to all denizens of the biosphere.

In the 1940s a new science appeared: "cybernetics".[1] One of its inventors, Norbert Wiener,[2] entitled his classical book on the subject *Cybernetics or control and communication in the animal and the machine.* There he wrote: "If the 17th and early 18th centuries are the age of clocks, and the later 18th and 19th centuries constitute the age of steam engines, the present time is the age of communication and control." Here communication—without which there can be no control—means transmission of information.

One of the essential tasks of 19th century science was the formulation of a theory of heat engines, a theory of heat. In the 20th century, it was a theory of communication—or information—that begged to be created (among others). And, surprisingly, it turned out that thermodynamics and information theory were connected.

[1] "Cybernetics is the interdisciplinary study of complex systems, especially communication processes, control mechanisms, and feedback principles." *Trans.*

[2] Norbert Wiener (1894–1964), American theoretical and applied mathematician.

The chief aim of information theory[3] consists first in clarifying the concept of information and its means of communication, and then discovering conditions for optimal communication, that is, optimal transmission of information. Thus we must first formulate a precise definition of "information" that reflects its everyday sense, while having a precise quantitative character.

We begin with some elementary examples. If we toss a coin, then the result—heads or tails—represents the communication of a definite amount of information about the toss. If the roll of a die results in a three, then this also constitutes information. The crucial question to ask here is: In which of these two situations do we obtain more information—in tossing a coin or rolling a die?

The obvious answer is: in rolling a die. For, in a coin-toss we have an experiment with just two equally likely possibilities, while in rolling a die there are six equiprobable possible outcomes. The probability of obtaining heads is 1/2, while that of a die coming up three is 1/6.[4] The realization of a less likely event represents greater information. Or, equivalently, the more uncertain an event is prior to receiving information about it, the greater the weight or "quantity" of that information when received. We have thus arrived at the conclusion that a quantitative measure of information should somehow depend on the number P_0 of (equally likely) possibilities. In the case of a coin-toss $P_0 = 2$, and in that of the roll of a die $P_0 = 6$.

It is also intuitively clear that if we roll a die twice (or roll a pair of dice), the result represents twice as much information as the result of rolling it only once. We conclude that information obtained from a sequence of independent trials is additive. If on the first roll the three-spot came up, and on the second the five-spot, then in total this represents twice as much information as the result of the first roll by itself.[5] Similar conclusions apply if we roll two dice.

Thus a measure of information should be additive over a set of independent events. On the other hand the number of ways a set of independent events can occur is multiplicative: If we roll a die twice (or roll a pair of dice), then $P_0 = 6 \cdot 6 = 36$. In general, if we have two independent events which can occur in P_{01} and P_{02} ways respectively, then the number of ways both events can occur is

$$P_0 = P_{01}P_{02}, \tag{7.1}$$

while an—as yet hypothetical—quantitative measure $I(P_0)$ of information should satisfy

$$I(P_0) = I(P_{01}P_{02}) = I(P_{01}) + I(P_{02}). \tag{7.2}$$

It follows that the dependence of the quantity of information yielded by an event on the number of ways the event can occur—the dependence of I on P_0—must be

---

[3] Information theory was created by the American mathematicians Claude Elwood Shannon (1916–2001) and Warren Weaver (1894–1978). *Trans.*

[4] See the first section of Chapter 4, including the footnotes, for the definitions of the probabilistic terms that follow. *Trans.*

[5] That is, as the first roll by itself yields about the two rolls. *Trans.*

Table 5: The binary forms of the numbers from 0 to 32.

	$0 = 0$		
$1 = 1$	$9 = 1001$	$17 = 10001$	$25 = 11001$
$2 = 10$	$10 = 1010$	$18 = 10010$	$26 = 11010$
$3 = 11$	$11 = 1011$	$19 = 10011$	$27 = 11011$
$4 = 100$	$12 = 1100$	$20 = 10100$	$28 = 11100$
$5 = 101$	$13 = 1101$	$21 = 10101$	$29 = 11101$
$6 = 110$	$14 = 1110$	$22 = 10110$	$30 = 11110$
$7 = 111$	$15 = 1111$	$23 = 10111$	$31 = 11111$
$8 = 1000$	$16 = 10000$	$24 = 11000$	$32 = 100000$

logarithmic:

$$I = K \log P_0. \tag{7.3}$$

The base of the logarithm and the constant K are not determined by (7.1) and (7.2), so they may be chosen arbitrarily. The established convention in information theory has $K = 1$ and the logarithmic base equal to 2. Hence

$$I = \log_2 P_0. \tag{7.4}$$

Then the basic unit of information, called a *bit*, is that obtained from a coin-toss, where $P_0 = 2$:

$$\log_2 2 = 1 \text{ bit}.$$

Thus information is calculated in bits, that is, in binary digits. The binary system is widely used in the technology of cybernetics, in particular in digital computers, since breaking a computation down into a sequence of operations each involving just two possibilities—inclusion or exclusion, say—greatly simplifies the computation.

Every number can be represented in the binary system by a sequence of 0s and 1s. A table of the whole numbers from 0 to 32 in both decimal and binary notations is given above.

In binary notation the multiplication table is especially simple:

$$0 \cdot 0 = 0, \;\; 1 \cdot 0 = 0 \cdot 1 = 0, \;\; 1 \cdot 1 = 1.$$

The addition table is:

$$0 + 0 = 0, \;\; 1 + 0 = 0 + 1 = 1, \;\; 1 + 1 = 10.$$

How many bits does an arbitrary three-digit number have? Since there are 900 such numbers—from 100 to 999—we have

$$I = \log_2 900 \approx 9.82 \text{ bits}.$$

Another way of looking at this calculation is to observe that the first digit in such a number (in decimal notation) can be any of the nine digits $1, \ldots, 9$, while the second and third each take any of the ten values from 0 to 9, so that

$$I = \log_2 900 = \log_2 9 + 2\log_2 10 \approx 9.82 \text{ bits.}$$

Note that since

$$\log_2 10 \approx 3.32,$$

the decimal base is equivalent to 3.32 bits, so that binary notation uses on average 3.32 times as many digits as decimal notation.

Computing information in terms of bits amounts to encoding an answer to a question as a sequence of "yes"s or "no"s. In his book *A mathematical trilogy*, A. Rényi[6] describes the "Bar-Kokhba" game,[7] popular in Hungary, which purportedly originated as follows. In 132 A.D. the Jewish leader Simon bar Kokhba led a revolt of the Jews against their Roman overlords.[8] A scout sent to spy on the Roman camp was captured, and had his tongue torn out and hands cut off. Returned to his own camp, he was thus unable to communicate either orally or in writing. However, Simon bar Kokhba was able to extract important information from him by asking a series of yes-no questions, which the poor scout could answer by nodding his head appropriately.

Using this approach, we shall verify the following assertion of Hartley:[9]

If in a given set containing N elements, a particular element x is singled out, about which it is known in advance only that it belongs to the set, then the quantity of information required to find x is $\log_2 N$ bits.

Thus according to Hartley, in order to determine which number between 1 and 32 your partner in the game is thinking of, you need to ask only five questions, since $\log_2 32 = 5$. And indeed, proceeding as in the Bar-Kokhba game, the first question should be: "Is the number greater than 16, yes or no?" This reduces the set of possibilities by a half, to 16. You continue reducing by half in this way, reducing the number of possibilities to 8, then 4, then 2, and finally 1.

It is important to observe that by proceeding somewhat differently we can ask five questions all at once, that is, without needing to know the answer to any question in advance. First write the numbers from 1 to 32 in binary notation, using five digits for all of them; thus the list will start with 00001 and end with 11111 (see Table 5). Suppose your partner is thinking of 14, that is, 01110 in binary notation. You can then ask the five questions all together in the form: "Is it true that in binary notation the first digit of the number you have in mind is 1, and the second digit, ..., and the fifth digit?" The answer will be: "no, yes, yes, yes, no".

[6] Alfréd Rényi (1921–1970), Hungarian mathematician. Worked mainly in probability.

[7] Similar to "Twenty questions". *Trans.*

[8] As a result of which an independent Jewish state was established in Israel, reconquered by the Romans in 135 A.D. Bar Kokhba was thus the last king of Israel. *Trans.*

[9] Ralph Vinton Lyon Hartley (1888–1970), American researcher in electronics, and contributor to the foundations of information theory.

The formula (7.4) provides the basis for solving such "search problems". Here is another one. Again the approach is essentially that of Simon bar Kokhba.

Suppose that we have 27 coins of which just one is false, weighing less than the others. What is the least number of weighings on a balance needed to find the false one? Each weighing with an equal number of coins in the pans of the balance, yields the quantity of information $I = \log_2 3$, since there are exactly three possibilities: the pans balance, the left pan is lighter, or the right is lighter. On the other hand, we know from Hartley's statement that finding the false coin requires $\log_2 27$ bits of information. Hence the false coin can be found in z weighings provided only that

$$z \log_2 3 \geq \log_2 27 = 3 \log_2 3,$$

that is, provided $z \geq 3$. Therefore three weighings suffice. In the first weighing, one places nine coins in each pan, in the second three, and in the third one.

Of course, in playing the Bar-Kokhba game efficiently, one needs to know which questions to ask!

Instead of numbers, let's consider letters. In the Roman alphabet there are 26 letters, and in the Russian 33. How much information does a single letter of some text yield in these two cases? The natural answer is: A letter of a passage in English yields $\log_2 26 \approx 4.70$ bits of information, while a letter from a Russian text yields $\log_2 33 \approx 5.05$ bits. However, these answers make sense only under the rather crude assumption that every letter appears with the same probability.

Informational entropy

Letters do not occur with overall equal probability; in any language written using an alphabet, some letters will occur on average more frequently than others. This average frequency—which is the same as the probability of occurrence of a letter—reflects the structure of the language in question. The reader may know Conan Doyle's story *The adventure of the dancing men*, in which Sherlock Holmes uses the known average frequency of occurrence of letters in English texts to decipher a coded message. (Actually, this idea appeared in literature considerably earlier, in Edgar Allan Poe's *The gold bug*, written before Conan Doyle was born.)

In deriving formula (7.4) it was assumed that the individual outcomes of the experiment or situation under consideration were equiprobable. What if they have various probabilities?

Suppose we have a message consisting of N successive cells—a text made up of N letters. Suppose further that each cell can contain any of M different letters. (If the text is in English, then $M = 26$.) Suppose also that the message contains N_1 occurrences of the letter a, N_2 of the letter b, and so on, up to N_{26} occurrences of z. Clearly

$$N = \sum_{i=1}^{M} N_i. \tag{7.5}$$

If the text we are presented with is sufficiently long, we may assume that the probability p_i of the ith letter occurring in any cell of any text is approximately given by

$$p_i = \frac{N_i}{N}, \; i = 1, 2, \ldots, M. \tag{7.6}$$

Then

$$\sum_{i=1}^{M} p_i = 1, \tag{7.7}$$

as should be the case. The number of N-letter sequences altogether is

$$P = \frac{N!}{N_1! N_2! \cdots N_M!}. \tag{7.8}$$

(We encountered this formula earlier; see (4.5) and (4.16).) By (7.4) the quantity of information in a single such message is[10]

$$I = \log_2 P = \frac{\ln P}{\ln 2} = \frac{1}{\ln 2} \ln \frac{N!}{N_1! N_2! \cdots N_M!}. \tag{7.9}$$

Using the approximation of $N!$ and the $N_i!$ given by (4.10) (and assuming N and the N_i all large), we infer that

$$\begin{aligned} I &\approx \frac{1}{\ln 2}\left(N \ln N - \sum_{i=1}^{M} N_i \ln N_i\right) \\ &= -\frac{1}{\ln 2} N \sum_{i=1}^{M} p_i \ln p_i = -N \sum_{i=1}^{M} p_i \log_2 p_i \text{ bits.} \end{aligned} \tag{7.10}$$

Hence if $N = 1$, that is, in the case of a single letter, we have

$$I_1 = -\frac{1}{\ln 2} \sum_{i=1}^{M} p_i \ln p_i = -\sum_{i=1}^{M} p_i \log_2 p_i. \tag{7.11}$$

The quantity $-\sum_i p_i \log_2 p_i$ was called by Shannon, one of the founders of information theory and communication theory, the *entropy* of any message involving M symbols occurring with probabilities p_i. We shall see below that the formula (7.11) does indeed afford a measure of the degree of uncertainty associated with a random such text, thus justifying the name "entropy".

What value does the entropy take in the case of English? The frequencies, that is, the probabilities, of occurrence of the letters (and also a space) in English texts are given in Table 6 below.
Substitution of these values in Shannon's formula (7.11) yields

$$I_1 = -0.164 \log_2 0.164 - 0.106 \log_2 0.106 - \cdots - 0.001 \log_2 0.001 \approx 4.0 \text{ bits.}$$

[10] Assuming all such N-letter strings equally probable. *Trans.*

Table 6: Probabilities p_i of occurrence of letters (and space) in English texts.

space .164	s .055	m .021	v .008
e .106	h .053	w .020	k .007
t .078	r .052	f .019	j .001
a .068	d .036	g .017	x .001
o .065	l .035	y .017	q .001
i .058	c .024	p .017	z .001
n .058	u .024	b .012	

This number is appreciably smaller than $I_0 = \log_2 26 \approx 4.7$. The amount of information communicated by the letters has decreased since we have incorporated in the calculation previously ascertained information about the frequency of occurrence of the letters.

However, in a language there are always correlations between letters—definite frequencies of occurrence not just of individual letters but also of strings of two letters (bigrams), three letters (trigrams), four letters (quadrigrams), and so on. A linguistic text represents a complicated "Markov chain", since the probability of a given letter occurring at a given place in the text depends on the preceding letter.

It is appropriate to say something here about the theory of Markov chains, one of the greatest achievements of Russian science. This theory, in which connected, probabilistically dependent chains of events are studied, was founded by A.A. Markov.[11] The calculation of the probability that several events will all occur is often made easy by the fact of the events in question being independent.[12] For example, if we toss a fair coin, the probability of getting heads is 1/2 regardless of the outcome of the previous toss. However such independence is far from always being the case.

We choose a card at random from a deck of cards. The probability of the card being of a specified suit is 1/4. If we then return the card to the deck, the probability that the next card chosen will be of that suit remains 1/4. However, if we do not return our chosen card to the deck, the probability that the next card chosen will be of the given suit will depend on the suit of the first card. For instance if that suit was clubs, then there remain in the deck 13 cards of each of the suits hearts, diamonds, and spades, but only 12 clubs. Hence the probability that the second card is also clubs is now 12/51, which is less than 1/4, while for each of the other suits the probability that the second card is of that suit is now 13/51, which is greater than 1/4.

A *Markov chain* consists of a sequence of events, each of whose probabilities depends on the outcome of the preceding event, or, in the more complex case, on

[11] Andreĭ Andreevich Markov (1856–1922), Russian mathematician.

[12] So that the probability of the combined events is the product of their individual probabilities. *Trans.*

the outcomes of several of its predecessors.

Clearly, any linguistic text constitutes a complex Markov chain. A.A. Markov himself applied his theory to a probabilistic analysis of Pushkin's *Eugene Onegin* and Aksakov's *The childhood of Bagrov's grandson.*

Today the theory of Markov chains and Markov processes is applied very widely in theoretical physics, meteorology, statistical economics, and so on.

Plans for a typewriter of maximal efficiency would have to take into account the frequency of occurrence of the individual letters, bigrams, and even trigrams of the language. The keys of rarely used letters should be relegated to one or another side of the keyboard, and the keys of letters that often occur together should be adjacent. The implementation of these two obvious principles is enough to greatly speed up the typing process.

Here is an experiment.[13] Write the 26 letters of the English alphabet on 26 slips of paper, and place these, together with a blank slip, in a bag. Then take out a slip, write down the letter on the slip (or leave a space if the slip is the blank one), return the slip to the bag, give the bag a good shake, and iterate this procedure a largish number of times. You will in all probability obtain a random text exhibiting no correlations, such as:

QFEZRTGPI BWZSUYKVLCPFMTQA UHXPBDKRQOJWN

If you now take account of the frequency of occurrence of the letters (and space) by placing 1000 slips of paper in the bag, with 164 blank, 106 with the letter E, 78 with the letter T, and so on, till, finally, just one slip for each of the letters J, X, Q, and Z (see Table 6), then the result of drawing out slips, with replacement, and recording the results, is more likely to resemble a text formally:

ENHRI VTUXSMO EHDAKC OTESL TNE

If you next in some similar fashion also include information about the frequencies of bigrams, then your result might be something like:

NTRETI ERANAL ITRONDTIOR QUCOSAAL INESTH

Taking account also of the frequencies of occurrence of trigrams, might yield:

NDETHER MENSTHEL HASITS THER OFTNCE QUIEXT

And, finally, including in addition information about the frequencies of quadrigrams, we might obtain:

THENSION ISTERNALLY ATORENCE OPERT IFUL

[13] In the original, there follows a description of an actual experiment using Russian carried out by the mathematician P.L. Dobrushin, reproduced from the book *Probability and Information* by A.M. Yaglom and I.M. Yaglom (Moscow: Nauka, 1973). The made-up English version presented here represents an attempt to convey the sense of the original. *Trans.*

The more extensive the correlations taken into account, the more the "texts" resemble English formally—without, of course, acquiring sense. The Laputan academics in Jonathan Swift's *Gulliver's travels* confined themselves to generating texts like the first one above, where no correlations whatever have been taken into account, by turning a letter-wheel and copying the letters in the hope of obtaining a text of substance.

In the case of Russian, the values of informational entropy as the more extensive correlations are successively taken into account, are as follows:

I_0	I_1	I_2	I_3	
5.00	4.35	3.52	3.01	bits

For English, Shannon carried the estimates much further:

I_0	I_1	I_2	I_3	...	I_5	...	I_8	
4.76	4.03	3.32	3.10	...	2.1	...	1.9	bits

Human languages characteristically have a considerable amount of built-in redundancy of information, for instance in the sense that it is possible to read a sensible text even if several letters are lacking. This redundancy can be quantified as follows: It would appear that I_n tends to a limit I_∞ as $n \to \infty$, that is, as the scope of the correlations grows without bound. The *redundancy* R of the language in question is defined by

$$R = 1 - \frac{I_\infty}{I_0}. \tag{7.12}$$

Rather than going all the way to the limit, we can consider the *redundancy relative to correlations up to length* n of the language, that is,

$$R_n = 1 - \frac{I_n}{I_0}. \tag{7.13}$$

For Russian the first few values of R_n are as follows:

R_0	R_1	R_2	R_3
0	0.13	0.30	0.40

and for English:

R_0	R_1	R_2	R_3	...	R_5	...	R_8
0	0.15	0.30	0.35	...	0.56	...	0.60

Thus in English the redundancy is certainly greater than 60%. This means that one can understand an English text even if only 40% of the letters are legible—provided, of course, that these are not clustered together.

To illustrate this, we describe an episode from Chapter 2 of Jules Verne's novel *Captain Grant's children.*[14] The protagonists find a bottle floating on the sea containing texts in English, German, and French. These are somewhat smudged by sea-water, so that only the following fragments of words are legible:

[14] *Les enfants du Capitaine Grant*, Paris, 1868.

62 Bri gow sink stra skipp
Gr that monit of long
and ssistance lost

7 Juni Glas zwei atrosen
greus bringt ihnen

trois ats tannia gonie austral
abor contin pr cruel indi
jeté ongit 37.11° lat

Thanks to the redundancy in the three languages, the protagonists are able to reconstitute all of the message except for the longitude:

"On June 7, 1862 the three-masted vessel 'Britannia', out of Glasgow, is sinking off the coast of Patagonia in the southern hemisphere. Making for the shore, two sailors and Captain Grant are about to land on the continent, where they will be taken prisoner by cruel indians. They have thrown this document into the sea at longitude ... and latitude 37.11°. Bring assistance, or they are lost."

And with this their adventures begin.

A more extreme example of the redundancy of language is afforded by the scene from L.N. Tolstoĭ's *Anna Karenina* of a conversation between Levin and Kitty:[15]

"Wait a moment", he said, seating himself at the table, "There is something I have been wanting to ask you about for a long time".

He looked straight into her eyes, which showed affection but also alarm.

"Ask away, by all means".

"Look here", he said, and wrote down the following initial letters:

w, y, r: t, c, b, d, t, m, j, t, o, n?

These letters stood for: "When you replied: 'That cannot be', did that mean just then, or never?"

Kitty understands what Levin wanted to ask her, and their coded conversation then continues. Here practically everything is redundant since Levin and Kitty are full to overflowing with the necessary information on the subject obsessing them both.

It is claimed that Tolstoĭ took this episode in the novel from his own life, that is, from his courtship of Sofia Andreevna Bers.

On the other hand, redundancy is indispensable in information, since in transmitting a message by any means of communication, there will inevitably be

[15]At this juncture in *Anna Karenina*, Levin, a good man and efficient husbandman, has been encouraged to approach Kitty once again, after an earlier rebuff. This pair, Kitty and Levin, are contrasted in the novel with the central pair, the beautiful Anna and the dashing Vronskiĭ, whose extramarital affair leads to Anna's ostracism by polite society, and her ultimate demise. *Trans.*

"noise", that is, random disturbances of one kind or another interfering with the transmission. (This is the basis of the children's game involving a defective telephone.) A ship in difficulty transmits the message SOS over and over to increase the probability of its reception.

Information and entropy

It is not difficult to see that Shannon's formula (7.11),

$$I_1 = -\sum_{i=1}^{M} p_i \log_2 p_i,$$

at least has the right form as an expression of entropy. Recall from Chapter 4 that for an isolated system consisting of N molecules with N_i molecules in state i, $i = 1, 2, \dots, M$, the entropy is given by the formula (4.19):

$$S = k \ln P = k \left(N \ln N - \sum_{i=1}^{M} N_i \ln N_i \right).$$

Writing $p_i = N_i/N$, the probability that an arbitrary molecule is in the ith state, we then have

$$S = -kN \left(\ln N - \sum_{i=1}^{M} \frac{N_i}{N} \ln N_i \right) = -kN \sum_{i=1}^{M} p_i \ln p_i,$$

whence the entropy per molecule is

$$S_1 = -k \sum_{i=1}^{M} p_i \ln p_i. \tag{7.14}$$

This formal resemblance between the formulae for I_1 and S_1 is not accidental.

We have often spoken of entropy as being a measure of the degree of disorder of a system, or as a measure of the extent to which information about the system is unavailable.

It is impossible to obtain information about an isolated adiabatic system, since any instrument—a thermometer, for instance—brought into contact with the system violates its isolation. Similarly, in obtaining information about one part of an open system one inevitably increases the entropy of some other part of the system.

Consider the freezing of water in a vessel. Here the entropy of the water decreases and information increases: the molecules were randomly distributed in the liquid, but are now, in the ice, fixed at the vertices of the ice's crystalline lattice, so we have a much better idea as to where they are. However, in order to freeze the water, we needed a refrigerator, and the freezing process will cause its temperature and entropy to rise.[16] And, of course, the rise in the refrigerator's entropy[17] must—the second law cannot be violated!—*more* than balance the drop in the entropy of the water due to freezing.

Thus for every bit of information obtained there is a cost in entropy. Now the units of information—bits—are dimensionless, whereas entropy is measured in calories/degree K, or joules /degree K, or ergs/degree K. To adjust the units so that formulae (7.11) and (7.14) coincide, we need to multiply the quantity of information I_1 in (7.11) by $k \ln 2 \approx 10^{-23}$ joules/°K. This gives the entropic equivalent of a bit. We see that in thermodynamic units a bit is very cheap.

Let's now attempt to estimate the number of bits contained in the whole of human culture. We shall assume that during the whole of its existence, the human race has produced 10^8 books (a greatly exaggerated figure), and that each book consists of 25 pages of authorial output. A standard such page would contain about 40,000 symbols. We shall ignore redundancy and attribute to each symbol 5 bits. Hence, altogether we obtain

$$10^8 \cdot 25 \cdot 4 \cdot 10^4 \cdot 5 \text{ bits} = 5 \cdot 10^{14} \text{ bits},$$

which is equivalent to a reduction in entropy of $5 \cdot 10^{-9}$ joules/°K! No, human culture should not be measured in thermodynamic units. But does this mean that the equivalence of entropy and information makes no sense?

To be perturbed about the relative difference in magnitudes of two equivalent physical quantities is inappropriate. For example, Einstein's equation giving the energy equivalent of mass,

$$m = \frac{E}{c^2},$$

where c is the speed of light, is fundamental in explaining the production of atomic energy. However, the equivalence factor here, namely $c^{-2} \approx 10^{-17}$ sec^2/m^2 is also extremely small: a little mass is equivalent to a lot of energy.

The small value of a bit in thermodynamic units may be interpreted as signifying that a quantity of information is a small difference between two large quantities, namely the amount of entropy before and the amount after the information has been obtained. From our discussion of information so far, it is clear that this difference must be non-negative.

An increase in information about a system is equivalent to a decrease in its entropy (once the units of measurement have been adjusted to coincide). Furthermore, information always refers to the microstates of a statistical system; entropy measures the unavailability of such information. Any change in the indeterminacy of the microstates of a physical system in a given macrostate, characterized, say, by the volume V and the temperature T, results in a decrease of entropy, or, equivalently, to an increase of micro-information.

[16] The compressor at the back of the refrigerator will heat up. *Trans.*

[17] Including that of the air immediately surrounding the refrigerator. *Trans.*

We shall now prove this assertion. If the system goes over from a distribution of probabilities $p_1, p_2, \ldots, p_M$ for its microstates, to a distribution $q_1, q_2, \ldots, q_M$, how does the amount of information change?

From formula (7.4) it follows that the change in information about an event (made up of equiprobable outcomes) resulting from a change in the probability of the event from p to q is

$$\Delta I = \log_2 q - \log_2 p = \frac{\ln(q/p)}{\ln 2}.$$

The expected value of the overall change in information due to the change $p_i \to q_i$, $i = 1, 2, \ldots, M$, is the sum of the partial changes, each multiplied by the corresponding probability q_i. Hence

$$\Delta I = \sum_i q_i \Delta I_i = \frac{1}{\ln 2} \sum_i q_i \ln \frac{q_i}{p_i}.$$

This quantity is positive unless $q_i = p_i$ for all i. This follows from the fact that for all $x > 0$ except $x = 1$,

$$\ln x > 1 - \frac{1}{x}.$$

For, provided $q_i \neq p_i$ for at least one i, this implies that

$$\sum_i q_i \ln \frac{q_i}{p_i} > \sum_i q_i \left(1 - \frac{q_i}{p_i}\right) = \sum_i q_i - \sum_i p_i = 0.$$

Thus a nontrivial change in the probabilities p_i always results in an increase in information.

The complementarity of information and entropy is illustrated by the evaporation of a liquid: information about the positions of the molecules is lost, namely their location in a circumscribed portion of space—the containing vessel—,while entropy increases by an equivalent amount. The information that the molecules of liquid were located in the vessel is transformed into "negative" information about the correlations between their positions and velocities in the vapor, affected by their collisions with each other. But this *lack* of information is just entropy.

Shannon's formula (7.11) for informational entropy, and the equivalent formula for thermodynamic entropy (7.14), share certain peculiar features. Entropy is defined in terms of a collection of random quantities, namely the values $E^{(i)}$ of the energy of the microsystems comprising the system in question. We wish to consider the extreme cases of such a system.

A state of the system will be one of zero indeterminacy, that is, will be fully determined, if $p_j = 1$ for some j, and $p_i = 0$ for all $i \neq j$. According to Nernst's theorem, this is the situation at absolute zero: $S(0°\text{K})=0$ (see the final section of Chapter 4).

The state of the system will have maximal indeterminacy, that is will satisfy $S = S_{max}, I = 0$, if all p_i are equal, that is, if the states are equally likely. This can be shown as follows. Since the probabilities p_i sum to 1, it follows that $p_i = 1/M$ for all i. We showed above that $\Delta I > 0$ for any change in the probabilities p_i to new probabilities q_i. In the case where $p_i = 1/M$ for $i = 1, 2, \ldots, M$, we have (see above)

$$\Delta I = \sum_i q_i \Delta I_i = \frac{1}{\ln 2} \sum_i q_i \ln \frac{q_i}{M^{-1}} = \frac{1}{\ln 2} \left(\ln M + \sum_i q_i \ln q_i \right).$$

Since $\Delta I > 0$ and the q_i are arbitrary, it follows that

$$-\sum_{i=1}^{M} \frac{1}{M} \ln \frac{1}{M} = \ln M > -\sum_i p_i \ln p_i,$$

for any probability distribution p_i other than the extreme one with all p_i equal, that is, entropy is largest in this case.

In contrast to microscopic information, the "price paid" in entropy for information about macrostates is high, that is, in an amount far from equivalent to the gain in information: the growth in entropy is many times greater than the quantity of information obtained. For example, when we toss a coin, we obtain one bit of macroscopic information, but the production of entropy resulting from the work of the muscles used in tossing the coin and from its impact with the floor, is many times larger than 10^{-16} ergs/°K (the thermodynamic equivalent of a bit), even if the coin is a small one. The larger the coin, the greater the discrepancy. This indicates why it makes no sense to estimate the information contained in books in calories per degree.

Maxwell's demon

The struggle to subvert the second law of thermodynamics, to prevent the growth in entropy, began a long time ago. To this end physicists recruited demons. We met with Laplace's demon in Chapter 5. In 1871, Maxwell introduced *his* demon to physics, in the form of "a being whose faculties are so sharpened that he can follow every molecule in its course, such a being, whose attributes are as essentially finite as our own, would be able to do what is impossible to us. For we have seen that molecules in a vessel full of air at uniform temperature are moving with velocities by no means uniform, though the mean velocity of any number of them, arbitrarily selected, is almost uniform. Now let us suppose that such a vessel is divided into two portions, A and B, by a division in which there is a small hole, and that a being who can see the individual molecules, opens and closes this hole, so as to allow only the swifter molecules to pass from A to B, and only the slower molecules to pass from B to A. He will thus, without expenditure of work, raise the temperature of B and lower that of A, in contradiction to the second law of thermodynamics".[18]

[18] James Clerk Maxwell, *Theory of heat*, London, 1872.

An analysis of this activity of Maxwell's demon was published by L. Brillouin[19] in 1951, expanding on work of Leó Szilárd[20] done in 1929, showing that the resolution to the paradox lies in information theory.

Brillouin's analysis[21] brings out the close connection between entropy and information especially clearly. The first question to ask is: What is required for the demon to be able to see the individual molecules?

If the system is isolated and in equilibrium at the constant temperature T_0, then the demon will have to be at that temperature also. Under these conditions any radiation will be black-body radiation, which, although he can observe it, will not enable him to see an individual molecule or determine its velocity. For that, the molecule would first have to be illuminated, so we would need to supply the demon with a battery-powered flashlight, whose filament would then need to be heated to a temperature T_1 exceeding T_0; we may assume, in fact, that $T_1 \gg T_0$. This condition is necessary to obtain visible light that can be distinguished from the background of black-body radiation in the vessel at temperature T_0, and ensure the condition that one quantum $h\nu$ of it[22] is much greater than kT_0, which is of the order of the thermal energy of a molecule. Neglecting the battery, we have that if E is the energy radiated per unit time by the filament of the bulb, then it loses entropy at the rate

$$S_{fil} = -\frac{E}{T_1}. \tag{7.15}$$

Before the demon intervenes, the energy E is absorbed by the gas at temperature T_0, as a result of which the entropy of the gas increases by the amount

$$S = \frac{E}{T_0} + S_{fil} = \frac{E}{T_0} - \frac{E}{T_1} > 0. \tag{7.16}$$

The demon will be able to see the molecule provided that at least one quantum $h\nu$ of light is scattered by the molecule to the demon's eye. The absorption of this quantum of energy raises the demon's entropy by the amount

$$\Delta S_{demon} = \frac{h\nu}{T_0}. \tag{7.17}$$

The demon's purpose is to use the information he obtains to decrease the entropy of the system. Now the initial entropy of the system was

$$S_0 = k \ln P_0. \tag{7.18}$$

[19] Léon Brillouin (1889–1969), French/American physicist.

[20] Leó Szilárd (1898–1964), Hungarian/American physicist. It was Szilárd who, in 1939, prevailed on Einstein to cosign the famous letter to President Franklin D. Roosevelt pointing out the feasibility of nuclear weapons and encouraging the initiation of a program to develop them ahead of Hitler's Germany—an initiative leading eventually to the Manhattan project.

[21] L. Brillouin, "Maxwell's demon cannot operate: information and entropy." J. Appl. Phys. 22 (1951), pp. 334–337. Later work on this theme, for example by Charles Bennett, may be found in the collection *Maxwell's demon 2: Classical and quantum information, computing*, H.S. Leff and A.F. Rex (Eds.), 2002. *Trans.*

[22] According to quantum mechanics, light of frequency ν comes in packets of energy of minimum size $h\nu$, where h is Planck's constant. *Trans.*

After receipt of the information, the indeterminacy of the system decreases, so that the statistical weight of the system is reduced from P_0 to $P_1 = P_0 - p$, say. This entails a change in the system's entropy by the amount

$$\Delta S_{info} = S - S_0 = -k(\ln(P_0 - p) - \ln P_0).$$

Since for $p \ll P_0$ (which holds in all practical cases),

$$\ln(P_0 - p) = \ln P_0 + \ln\left(1 - \frac{p}{P_0}\right) \approx \ln P_0 - \frac{p}{P_0},$$

we have, approximately,

$$\Delta S_{info} = -\frac{kp}{P_0}. \tag{7.19}$$

Hence the total change in entropy is

$$\Delta S = \Delta S_{demon} + \Delta S_{info} \approx k\left(\frac{h\nu}{kT_0} - \frac{p}{P_0}\right) > 0, \tag{7.20}$$

since $h\nu > kT_0$ and $p \ll P_0$. We conclude that, after all, the entropy of the system increases just as the second law says it should. The demon was not able to violate that law. Brillouin says: "All the demon can do is recuperate a small part of the entropy and use the information to decrease the degradation of energy".

To summarize: At the first stage of the demonic process, entropy increased by the amount ΔS_{demon}, at the second some of this entropy was transformed into information, and, finally, this was then used to partially decrease the entropy.

The efficiency of the work of the demon is given by the ratio of the decrease in entropy resulting from the information he received, to the total increase in entropy:

$$\eta = \frac{|\Delta S_{info}|}{\Delta S} = \frac{p/P_0}{h\nu/kT_0 - p/P_0} \ll 1. \tag{7.21}$$

Brillouin, and also the Soviet physicist R. P. Poplavskiĭ, were able to show that η depends on the relative decrease in temperature achieved by the demon, that is, on

$$\frac{\Delta T}{T_0} = \frac{T_B - T_A}{T_0} = \theta,$$

where $T_B - T_A = \Delta T$ is the temperature difference achieved by the demon between the two halves of the vessel. Thus for instance if $\theta \ll 1$, then the coefficient of useful action in the corresponding Carnot cycle (see (1.25)) would be

$$\begin{aligned}\eta_C = \frac{T_B - T_A}{T_B} &= \frac{(T_0 + \Delta T/2) - (T_0 - \Delta T/2)}{T_0 + \Delta T/2} \\ &= \frac{\Delta T}{T_0 + \Delta T/2} = \frac{\theta}{1 + \theta/2} \approx \theta. \end{aligned} \tag{7.22}$$

Hence for $\theta \ll 1$ both coefficients of useful action η and η_K are essentially equal. The coefficient η measures the degree of irreversibility of a process in which order is created, while the Carnot coefficient η_C measures the degree to which heat can be reversibly transformed into work. However, as Poplavskiĭ stresses, to obtain work two stages are necessary: an informational one, that is, a control stage, and a thermodynamic one.

Maxwell's demon, like Laplace's, has been the subject of poetry. In Andreĭ Belyĭ's greatest poetical work "First encounter", one finds the following verses:

With austere physics my mind
Was overfilled by: Professor Braine.
With neck bent back, and ruffled mane,
He sang of cosmic gloom,
And of how Maxwell annihilated entropy
With his paradoxes.

Poetry is not so far from science as those people who so rigorously separate physicists from lyrical poets like to think (see the last section of Chapter 8). Some lines later in the poem, Andreĭ Belyĭ utters a remarkable poetic surmise:

In the Curies' experiments the world burst forth
With the exploding atom bomb
Into electronic streamings
Like a disembodied hecatomb.

This poem was published in 1921.

Obtaining and creating information

We have thus convinced ourselves that one must pay for information with an increase in entropy. In macroscopic processes this cost can be considerable. The entropic equivalent $k \ln 2$ that we obtained (see above) for a single bit of information is just the lower limit of this cost.

We shall now calculate the cost of information in units of energy.

Suppose we have a quantity of an ideal gas at pressure p and temperature T, consisting of N molecules, and that as the result of fluctuations the volume has decreased from V to $V - \Delta V$. The work done in achieving this is $W = p\Delta V$ (see Chapter 1). The information obtained thereby is calculated as follows: Each molecule was formerly contained in the region of volume V with probability 1, that is, with certainty. The probability that it was formerly in the now contracted region of volume $V - \Delta V$ is clearly $1 - \Delta V/V$. The probability that all N molecules were in that subregion of the original region is therefore $(1 - \Delta V/V)^N$. As a result of the compression of the gas, the N molecules occupy the smaller region, so that we now have greater information about them, namely in the amount

$$\Delta I = \frac{1}{\ln 2} \ln \frac{1}{(1 - \Delta V/V)^N} = -\frac{1}{\ln 2} \ln \left(1 - \frac{\Delta V}{V}\right)^N \approx \frac{1}{\ln 2} N \frac{\Delta V}{V}. \tag{7.23}$$

Hence

$$\frac{W}{\Delta I} \approx \frac{p\Delta V}{\frac{1}{\ln 2} N \frac{\Delta V}{V}} = \frac{pV}{N} \ln 2 = kT \ln 2. \tag{7.24}$$

This represents the work done per bit of information obtained, that is, the quantity $kT \ln 2$ is the minimal cost in energy of one bit of information. At $T = 300°$K (room temperature), we have $kT \ln 2 = 2 \cdot 10^{-21}$ joules.

Every physical measurement yields information, and therefore entails a loss in energy as well as a gain in entropy. Measuring procedures—that is, processes for gathering information about the macrostate of a system and creating order in the system—are irreversible in principle. The study of the thermodynamics of such informational processes has become of crucial importance in physics in view of the fact that the costs in energy and entropy rise with the precision of physical measurements.

However, physics is not limited to the taking of measurements. The results of measurement are transmitted and used, that is, the information obtained through measurement is processed (in part on computers). And this too must be paid for.

Thus we see that the concept "information" has a well-defined thermodynamic sense. This being so, one can safely ignore the claim sometimes found in print to the effect that the concept of information is vague and non-physical.

In previous sections we spoke in passing of obtaining or receiving information, without mentioning the significance of such processes. We now fill the breach. Note first that the concept of information is characterized by the following two postulates:

1. Information signifies a choice of certain events from a large collection of events (equiprobable or otherwise).

2. Such choices as in 1 are considered information only if they can be received and remembered.

In the development of information theory in the works of Wiener, Shannon, and others, where the aim was that of solving problems in communications, more attention was paid to processes of transmission of information than to those of reception. In the standard theory the receptor has very limited capabilities: all it can do is distinguish one letter from another, one coded symbol from another.

The situation is quite different in informational physics and informational biology (to be discussed in the next chapter). There it becomes especially important to investigate how information is *received*.

We can begin our discussion of the reception of information by considering our everyday experience of it. What sort of process *is* it?

First, it is clear that reception of information is an irreversible process (paid for by an increase in entropy!). Information may be forgotten or wasted by the receptor in some other way, but cannot be recouped.

Second, reception is a non-equilibrial process. If a source of information and its receptor are in equilibrium, an exchange of information is taking place between them—both direct and in reverse—, and these flows must balance one another.

Third, since reception of information indicates the creation of order in the receptor system (I read a poem and remember it!), this is not just a non-equilibrial process, but one far from equilibrium. A receptor system is a dissipative system.

Fourth, for the reception of information it is necessary that the receptor possess a certain level of capability, of capacity to take in the information. However, although necessary, the presence of the requisite capacity is not sufficient for reception to occur. Before me there is an anthology of poetry in a language I am not familiar with. In this case, I do not have the necessary level of receptor capacity (that is, the preliminary store of information that might, for instance, result from the study of the language in question) and so am not in a state to receive the information contained in the anthology. On the other hand, if it is an anthology of Russian poems, then, since Russian is my native tongue, I do possess the requisite level of receptor capacity (including the linguistic preparation required for an aesthetic appreciation). However, I do not wish to read poetry just now, having other things in mind.

Hence the fifth requirement for the reception of information, namely that there be an element of purpose, an aim. The presence of purpose indicates instability, since the realization of an aim represents a transition from a less stable state to a more stable one.

Thus reception of information is an irreversible process of transition of a dissipative system from a less stable state to a more stable one. As always in such situations (see Chapter 6) such a process must involve the export of entropy from the receptor system.

Of course, information is considered as having been received only if it is remembered for a longer or shorter period. Reception and remembering of information go together indissolubly.

Information can be accumulated and stored. Books, or, to take a very different example, the genomes of all species on earth, represent stores of information.

We receive and remember only macroscopic information. This kind of information demands payment not in equivalent amounts of entropy, but in significantly larger amounts. That is why the estimate we made earlier of the equivalent in entropic units of the information contained in all the books ever written is really meaningless.

How is information generated? An answer to this question has been given by the theoretician A. Koestler,[23] namely, that information is created through the

[23]This *may* be Author Koestler (1905–1983), Jewish-Hungarian/British polymath writer. Another possibility is: G. Kastler, author of *The origin of biological organization*, 1967. *Trans.*

committing to memory of a random choice. He gives the following illustration: Suppose I place my suitcase in a cloakroom locker at the railway station, and encode the combination lock with a four-digit combination, knowledge of which will enable me to open the locker. By committing to memory (or writing down) that randomly chosen sequence of four digits I have created information, namely in the amount $\log_2 9000 = 13.13$ bits.

The creation of new information always proceeds in jumps rather than gradually. Although essentially non-equilibrial, it bears similarities to phase transition. However, ordinary equilibrial phase transitions such as the crystallization of a liquid, do not involve the generation of any new information; there is no choice involved, everything proceeds in prescribed fashion according to the appropriate laws of physics. On the other hand, the creation of an actual crystal with peculiar flaws in its lattice—fissures and impurities, for instance—may be viewed as representing a random choice, and thus as giving rise to new information.

We have introduced the reader to two demons—Laplace's and Maxwell's—but to only one "billiard" (see the second-last section of Chapter 5). Here is another, the so-called "Chinese billiard".[24] There is a board, or field of play, on which there are distributed fixed studs or pins and shallow pits or wells. Next to each well there is a number indicating the score for that well. Metal balls are fired along the board, one at a time. After a succession of collisions with the pins—so that its trajectory is usually quite chaotic—a ball ultimately comes to rest in one of the wells. The motion of a ball till it comes to a halt in a well provides a model of entropy. The coming to rest in a well represents the achieving of a relatively stable state as the result of a random choice from among a multitude of possibilities, and therefore models the generation of information. The cost in entropy for this information is huge. For, if there are 32, say, wells on the board, then the amount of information generated (and received) is 5 bits, while the amount of heat per unit temperature generated by the initial propulsion and subsequent collisions of the ball up to the time it comes to rest exceeds $5k \ln 2 = 5 \cdot 10^{-23}$ joules/°K many times over.

It should be emphasized that at the present time the creation of a genuine physical theory of reception, remembering (or recording), and generation of information, that is, of these specific irreversible processes in appropriate dissipative systems, is encountering significant difficulties.

The value of information

Every communication of information has a definite content and meaning, and is of some value or other to the recipient. Standard information theory—which is what we have been concerned with up till now—completely neglects the question of the content of information. However, this is actually a virtue rather than a defect of that theory; clearly, if the main concern is the transmissive capability of

[24]The basic form of "pinball". *Trans.*

a communications channel—for example, an electric telegraph system—then it is not at all appropriate to take into consideration the content of the transmitted telegrams.

On the other hand, there are many scientific problems having to do directly with the value of certain information. These are mostly problems arising in biology and various aspects of the humanities, and we will be discussing them to some extent in the next chapter.

The question of the value of information has been investigated by a number of Soviet scholars, including M.M. Bongard, R.L. Stratonovich, and A.A. Kharkevich. Chapter VII of Mikhail Moĭseevich Bongard's very interesting book *The recognition problem*[25] (Moscow: Nauka, 1967) is entitled "Useful information". Bongard links the degree of usefulness of a message, that is, the value of the information it contains, with the increase in the probability of achieving some objective as a result of receiving the message. Thus the value of the information contained in the message is given by the formula

$$V = \log_2 \left(\frac{p'}{p} \right), \tag{7.25}$$

where p and p' are respectively the probabilities of achieving the relevant aim before and after receipt of the information. The notions of value introduced by Stratonovich and Kharkevich in their works are similar.

It is clear that the value of information cannot be defined independently of its reception, since we can judge the value of a message only from the consequences of its apprehension by the receptor. Thus, in contrast with the definitions of *quantity* of information expressed by Hartley's formula (7.4) or Shannon's (7.11), it is impossible to give a universal definition of the *value* of information. The value is only revealed upon reception, and is intimately connected with the level of that reception. Here is a simple example: Consider Volume 2 of V.I. Smirnov's *A course in higher mathematics*. This book contains a wealth of information. But what is its value?

The natural response is: For whom? For a preschooler the answer is zero, since he or she lacks the necessary preparation, hence does not possess an adequate level of reception, and is therefore not in a position or state to apprehend the information contained in the book. For a competent mathematics professor the value is again zero since he knows the contents of the book very well. The information in question is of greatest value for students taking the course for which this textbook was written—since this textbook is indeed an excellent one. The graph of the value of the information against the level of preparation—which might be called the "thesaurus" from the Latin for treasure or store—here passes through its maximum (Figure 7.1).

Clearly, the specific value given by formula (7.25) represents a characteristic chiefly of the reception of the information. The increase in the probability of

[25] An English version was published in 1968. *Trans.*

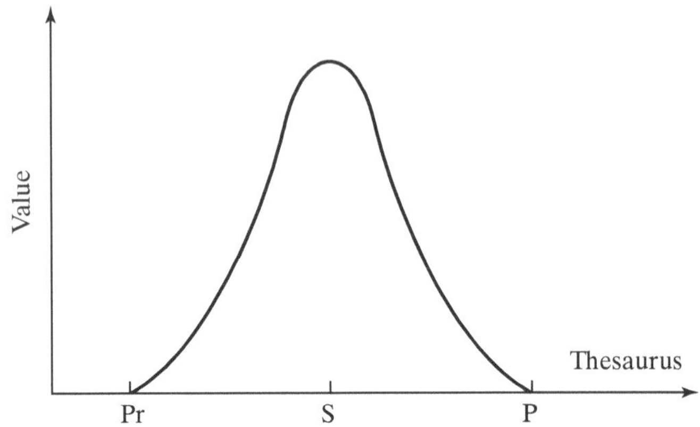

Figure 7.1: Graph showing the dependence of the value of the information on the "thesaurus": Pr is for preschooler, S for student, and P for professor.

achievement of the relevant aim is determined by the receptor whose aim it is. Of course, the concept of aim or purpose, though highly subjective when applied to a human receptor, is entirely objective when applied to phenomena of physics, chemistry, and biology. As mentioned earlier, the presence of an "aim" in connection with such phenomena signifies merely a definite instability. The "aim" of genes for structure is the synthesis of albumins (see the third section of Chapter 8). In this sense of the word, the concept of aim or purpose does not go beyond the bounds of physics or chemistry.

According to formula (7.25), redundant or repeated information has zero value since it does not result in any change in the probability of achieving the aim. Note also that the value of the information received can be negative if it is actually disinformation, that is, false information making it harder to realize the aim.

Practical application of formula (7.25) presents difficulties. Let us try to simplify the problem a little. Starting from the fact that redundant information has zero value (for example, repetition of the same facts conveys no new information), we shall redefine the value of information as the degree of irredundancy, that is, the degree of irreplaceability of the communication as a whole or else of a single element of it, for example, a symbol of code.

Suppose the message contains N_0 such symbols or letters, and the total amount of information is $N_0 I_0$. At the next level of reception, where the frequencies of occurrence of letters is taken into consideration (see the second section of this chapter), the amount of information conveyed reduces to $N_0 I_1$ where $I_1 < I_0$. (As we saw earlier, for Russian, $I_0 = 5.05, I_1 = 4.35$ bits.) We now reduce the number of letters to N_1 in proportion to the reduction in information; that is, we define N_1 by (see the second section above)

$$\frac{N_1}{N_0} = \frac{I_1}{I_0} = 1 - R_1.$$

Thus, we have, in effect, discarded the redundant letters. As a result, the remaining letters have acquired greater value, namely I_0/I_1, by virtue of being less redundant. Successively taking into account frequencies of occurrence of digrams, trigrams, and so on, we obtain the increasing sequence of values

$$1, \quad \frac{I_0}{I_1}, \quad \frac{I_0}{I_2}, \quad \frac{I_0}{I_3}, \ldots, \frac{I_0}{I_n},$$

which for Russian are 1, 1.15, 1.42, 1.66, ... And what is the goal here? It is that of obtaining precise, unambiguous information freed of all redundancy.[26]

This approach to defining the value of information turns out to be useful in connection with various informational processes—primarily biological—to be discussed in the next chapter. There we shall also consider informational aspects of artistic creation and its reception. These extended notions of information theory, which formerly allowed of rigorous development only as a theory of communications, in which reception was ignored, turn out to be of universal application. Merely applying the terminology reveals a great deal.

There are situations where the value of a piece of information, that is, the weight of the consequences of obtaining the information, is incomparably greater than the actual quantity of the information. For example, a win at the roulette wheel, or the result of rolling a die, may be of tremendous significance for the gambler. And the receipt of a single bit of information by drivers suffices to change the flow of a stream of traffic—a traffic light's changing from green to red. Situations such as these are "trigger" situations, depending on the processing of a large store of preliminary information. The conversation between Kitty and Levin quoted above (see the second section), mysterious to outside observers, is of a similar character. Trigger situations are constantly being realized in biological systems.

[26] And so of maximal value. *Trans.*

Chapter 8

Entropy, information, life

To bring joy to the gods and to men,
Apollo, guide to the muses,
Led nine radiant maids from crest to crest.
Euterpe sang a sweet song to the enamoured,
And for these also Erato composed verses, plucking a zither.
Calliope, Urania, and Clio brought men knowledge.
With learning came wise Polyhymnia's poems,
And Terpsichore revealed the deep essence of dance.
Euripides was betrothed to stern Melpomene,
While Aristophanes studied at Thalia's feet alone.
But the muses' guide did not know that all nine
Were mere incarnations of the main muse:
The tenth, named Entropia.

The thermodynamics of living organisms

It has become a commonplace to hear organic life spoken of as being "anti-entropic". What those who use this expression have in mind is the observed growth of order in the course of both individual and overall evolutionary development. In this respect a living organism is entirely unlike a system that, in accordance with the second law of thermodynamics, tends to a state of maximal entropy, of maximal disorder. This striking disparity has served as the inspiration for a number of publications rejecting the applicability of the second law to living systems. However, such publications are unscientific.

We have already seen what is at issue here. A living organism is an open system far from equilibrium. In analysing the balance of entropy of an astronaut in a space ship (see Chapter 6), we convinced ourselves that all is as it should be with regard to thermodynamics. Both the development and the stationary state of an organism maintain themselves by the export of entropy into the surrounding medium. When a system is far from equilibrium, there may arise a kind of order peculiar to dissipative systems, as the result of an increase in fluctuations up to the macroscopic level (see the fourth section of Chapter 6).

The French scientist J. Monod,[1] one of the founders of molecular biology, wrote that a living organism differs from a dead one of the same mass by possessing greater order. He was mistaken. The order of an organism in the sense of a regular arrangement of atoms and molecules, is significantly less than such order in a crystal—if only because an organism contains a large proportion of liquid. The types of order manifested in an organism and a crystal are fundamentally dissimilar: A crystal is ordered equilibrially, while an organism is ordered as a dissipative system far from equilibrium. In the book by Schrödinger cited in the third section of Chapter 6, a living organism is called an "aperiodic crystal". Although this phrase is pregnant with meaning, we shall postpone elaborating on it till later. Instead, we turn to something else of significance from that book.

Schrödinger asks the reader: "Why are atoms small?"—and immediately points out that the question is incorrectly posed. In physics, talk of the large and the small is out of place, since there only relative size is important: in physics one compares. Atoms are small—compared with what? Compared with our usual measures of length, a foot or a meter. And these measures were determined originally by the dimensions of the human body. Thus atoms are much smaller than living organisms. Hence the question should be reformulated as follows: "Why is an organism made up of a large number of atoms?" And this question is legitimate. For example, the tiny bacterium *Mycoplasma laidlawii* consists of about 10^9 atoms.

The crucial point here is that exact physical laws are required for the functioning of an organism, while, on the other hand, all the physical and chemical laws actually used to explain the existence and functioning of organisms are based on statistics and therefore only approximate. (We are not referring here to the quantum laws, which determine the structure of the molecules of which organisms are composed.) The precision of statistical laws depends on the involvement of a large number of atoms. Thermodynamics, in particular the second law, do not apply to small collections of particles; any relative fluctuations would then be too large. In a statistical system with only a few basic elements, neither a stable equilibrial order nor a stable dissipative order could arise.[2] An organism is a dissipative system capable of macroscopic amplification of fluctuations; here the adjective "macroscopic" is crucial.

The first scientist to argue that the state of a living organism is fundamentally one of *non*-equilibrium, was the Soviet biologist E.S. Bauer. In 1935 he wrote: "...living systems are never in equilibrium and at the expense of their free energy

[1] Jacques Monod (1910–1976), French biologist. Nobel laureate in physiology or medicine in 1965.

[2] The emphasis here is on the word "statistical". A purely mechanical system with only a few basic elements can certainly attain equilibrium. *Trans.*

they work incessantly to thwart the equilibrium demanded by the laws of physics and chemistry under the prevailing external conditions."

To repeat: a living organism represents an open macroscopic system far from equilibrium, that is, a dissipative system. Although it is essential to be aware of this fact, it does not suffice by itself to explain the chief peculiarities of life in terms of thermodynamics and statistical physics.

A living organism is an extremely complicated chemical machine—chemical, since it functions under conditions of constant temperature and pressure. Hence the sources of the energy indispensable for carrying out the various kinds of work required by the organism for its functioning, are not thermal but chemical. This energy is stored primarily in the compound adenozin triphosphoric acid (ATP). When one of the three phosphate bonds present in this compound is broken, a quantity of free energy of the order of 40 kilojoules/mole is liberated. This energy is used to activate chemical processes—in particular those involved in the synthesis of albumins[3]—, for the transportation of various substances into cells and out of them, and for the harnessing of nerve impulses, which are electrical in nature, to do mechanical work. Life could not exist without motion. Motion inside cells, such as the movement of chromosomes when cells divide, movement of organs and the organism as a whole, muscular contraction: these all require the expenditure of the chemical energy stored in ATP. The storing itself is achieved via the breathing processes in animals and plants and photosynthesis in plants.

Living organisms differ from the machines—in particular, robots—that men have so far managed to build, not only in the nature of their energizing processes, but also in the nature of their communication processes. Direct communication and feedback loops used in the control of artificial machines employ electrical, magnetic, and mechanical means of transmission of signals. In a living organism, on the other hand, the transmission, coding, and reception of informational signals is in the last analysis always chemical: molecules and ions serve as the signals themselves, and molecular systems as sources, transformational devices, and receptors of the signals. For example, an albumin ferment catalyzing a certain biochemical reaction acts as a transformer of a signal in that it catalyzes the conversion of certain signal molecules into others.

As mentioned earlier, any kind of work can be carried out by a thermodynamic system provided it incorporates differences sufficing to determine the desired acting force. Thus in Carnot's cycle, as embodied by a heat engine, the determining factor is the temperature difference between heater and cooler. In a living organism, everything is determined by differences in concentration of chemical substances, by differences in chemical potentials. An organism is alive thanks to delicate and precise chemical balances.

Since an organism is an open system, there are many interdependent processes going on within it. The dissipation function (see the second section of Chapter 6) is the sum of the products of the "generalized forces" by the "generalized

[3]Certain water-soluble proteins responsible for the functioning of cells. *Trans.*

flows":

$$\sigma = \sum_i X_i J_i.$$

This explains why reactions occur in organisms that are not possible *in vitro*, that is, in a flask. Here is an example. An albumin is synthesized from amino acids, molecules of which have the structure

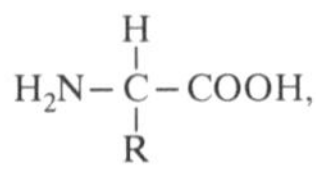

where R is any of several possible radicals. All naturally occurring albumins are constructed from the twenty amino acids differing in the radical R. An albumin chain is a kind of text written using an alphabet of twenty letters. The amino acids are combined into a chain by means of peptide bonds —CO—NH—. The bonding of each pair of amino acids:

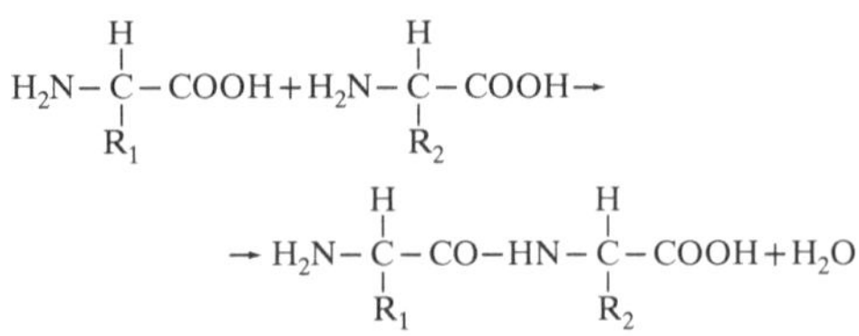

is accompanied by the secretion of a single molecule of water. The whole process takes place in a watery medium, namely in the cell's cytoplasm, where water is abundant. In such a medium the reaction could not proceed of its own accord since the peptide bonds would disengage sooner than connect pairs of amino acids. However, the process is accompanied by the dissociation of ATP, which furnishes the free energy essential to the synthesis of the albumin.

Albumins can, in fact, be synthesized *in vitro*, but by completely different means.

Combinations of chemical reactions with processes involving transportation of substances—with diffusion processes—play an especially important role in life processes. The energy liberated from ATP is used to produce differences in concentrations of ions between the interior and exterior of the cell. In a red blood cell,[4] for example, the concentration of potassium ions is much greater than in the surrounding plasma, while for sodium ions it is the other way around. This active transportation of substances through the cell membrane—the export of sodium and the import of potassium—takes place as a result of hydrolytic splitting of ATP.

In a heat engine there is a clear separation between the mechanical and statistical parts, between the mechanical and statistical degrees of freedom.[5] The

[4] Or *erythrocyte*, meaning red sphere.

[5] Roughly speaking, the "degrees of freedom" of a system refer to the (number of) parameters needed to determine it. *Trans.*

mechanical part consists of a metallic cylinder and piston, and the statistical part consists of the gas expanding and contracting, and doing the work. The entropy of the mechanical part of the engine is essentially unchanging, but that of the working material—steam, say—does change, since that part constitutes a statistical system subject to the laws of probability.

In a living organism the situation is far more complicated. Here the "mechanical" and "statistical" degrees of freedom (the quotation marks are used with good reason) are intimately combined; they cannot be separated from one another even by close scientific scrutiny. Take, for example, a muscle. Here the active substances (or working materials), namely the contracting albumins actin and myosin, also comprise the whole structure of the system. Thus the situation is very different from that of a heat engine, where the working material—steam or some other gas—is located within a mechanical structure.

Today we largely understand the physical basis of life, the physical fundamentals of the functioning of cells and organisms, although research on several of the main questions of theoretical biology and biophysics is still at the initial stage. At least we understand that life is not "anti-entropic", a word that is bereft of meaning. On the contrary, life exists because there *is* entropy, the export of which supports biological processes at all levels—from cells to the biosphere as a whole.

Thermodynamics has always been associated with the life sciences. We have seen (in Chapter 2) how Mayer and Helmholtz arrived at the first law from observations of biological phenomena. The second stage in the development of thermodynamics—the construction of a linear thermodynamics of open systems—saw associated processes in a system explained, and the achievement of an understanding of stationary, though nonequilibrial, states of an open system. Finally, at the third stage—that of synergetics, the physics of dissipative systems—physics united with biology to reveal the meaning of biological order and biological development.

Today we can say with certainty that none of the biological phenomena that have been investigated so far contradicts the established laws of physics. Of course, it is sometimes necessary to introduce new concepts, but these do not affect the underlying physical theory. This situation differs from that of physics itself: quantum theory and the theory of relativity arose because the earlier physics encountered limits to its applicability. No such limits to the applicability of the fundamental laws of physics have turned up in biology.

Modern physics—in particular, all those parts of physics relating to entropy—work splendidly in biology, and there is no need to introduce such concepts as "biofields" or "bioplasmas". There are no such things in nature.

Biological evolution, entropy, and information

The Earth's biosphere is a system with a long past, first having undergone change during a long prebiological phase culminating in the appearance of life, and then during a long phase of biological evolution. This evolution has so far reached the point we find ourselves at, surrounded by a great variety of life forms. Life appears to be localized on Earth, an exceedingly minute portion of the universe. So far no evidence has been found of life anywhere else. Stories of creatures from other planets are unrelated to science. I. S. Shklovskiĭ advanced various arguments for the impossibility of our meeting with extraterrestrial life; however, none of them is rigorous or can even be made so.[6]

The period of time during which biological evolution has been going on is of the same general order as that for which the universe has existed, which is about $1.4 \cdot 10^{10}$ years. The Earth itself is estimated to be around $4.5 \cdot 10^9$ years old. Life is reckoned to have first appeared about $3.9 \cdot 10^9$ years ago; this estimate is based on the oldest fossilized imprints of primitive bacteria and algae—so-called "microbions"—so far discovered.

What does the evolution of the universe, solar system, and Earth, have in common with biological evolution? In all cases the relevant evolution involves the generation of new information, be it in the form of galaxies and stars, planets, the formation of mountains through foldings of the Earth's crust, or of biological species. As we have seen, new information is created as the result of an arbitrary choice, arising in turn from the instability of the original state of the system in question together with the availability of various more stable states, from amongst which the choice is made. The generation of new information—be it cosmological, geological, or biological—has the character of a phase transition: it is new order that arises not gradually but all at once—at a jump.

It has in fact been shown that the equations governing population genetics, which describe how new biological species arise, have the same form as those governing phase transitions.

There are striking similarities between cosmic development and biological evolution. The appearance of cosmic inhomogeneities—stars and galaxies—as a result of gravitational instability, proceeds by competition and natural selection: gravitational inhomogeneities compete with one another for the condensing matter.

The cosmic, geological, and biological evolutionary processes—processes of structure-formation, of the appearance of order out of initial chaos—all take place as a result of the export of entropy, its efflux into the surrounding medium.

What we have said here of the biosphere holds true also of an individual organism, in the sense that in developing from an initial fertilized ovum, say, it retains in its structure and the way it functions, a memory of the prior biological evolution that has led to that organism.[7]

---

[6] Iosif Samuĭlovich Shklovskiĭ (1916–1985), Soviet astronomer and astrophysicist. Since the present book first appeared, the search for extraterrestrial life has developed very considerably. See, for example, *The search for life in the universe*, by D. Goldsmith and T. Owen, University Science Books, 2002, or *An introduction to astrobiology*, edited by I. Gilmour and M. Stephton, Cambridge University Press, 2004. *Trans.*

[7] "Ontogeny recapitulates phylogeny." *Trans.*

One might include also the 20th century in the "age of Darwin", since it was an era dominated by the study of temporal processes, processes of development, of evolution, when problems related to stability were the chief concern of science.

Information is created each time a new organism comes into the world. In fact, sexual reproduction signifies the remembering of a random selection. There is, after all, no law prescribing which descendants of two particular people will emerge into being.

How much information is contained in an organism? L.A. Blumenfeld[8] made the following estimate for the human organism: The basic amount of information in a human organism is determined by the ordered arrangement of amino acids (or amino acid residues[9]) in 7 kilograms of albumins, which contains about $3 \cdot 10^{25}$ such residues. Since there are twenty sorts of amino-acid residues, each of them represents $\log_2 20 = 4.32$ bits of information. Multiplying by $3 \cdot 10^{25}$, we obtain altogether $13 \cdot 10^{26}$ bits. Contributions from other sources turn out to be significantly smaller. For example, deoxyribonucleic acid, DNA, represents a text written in a four-letter alphabet, since there are just four different kinds of links—called nucleotides—in the DNA chain. Hence each nucleotide represents $\log_2 4 = 2$ bits of information. Since a human organism contains about 150 grams of DNA, made up of around $1.5 \cdot 10^{23}$ nucleotides, we obtain, as the contribution from the organism's DNA, about $3 \cdot 10^{23}$ bits. The ordered arrangement of the roughly 10^{13} cells of the organism contributes a further $4 \cdot 10^{14}$ bits, and the ordered arrangement of the 10^8 macromolecules of biopolymers in each cell another $1.6 \cdot 10^9$ bits.

However, this calculation is of little value to biology. In the first place, it does not distinguish redundant from irredundant information. Incidentally, as already mentioned, an organism is an "aperiodic crystal", and as such must contain a great deal of irredundant, and hence valuable, information. This represents an important difference between an organism—an aperiodic crystal not in equilibrium—and an ordinary periodic crystal in equilibrium.

Furthermore, such calculations make various tacit assumptions. The information contained in an albumin molecule is not limited to its bare molecular structure, that is, to the sequence of amino-acid residues comprising it. As already mentioned (see the sixth section of Chapter 3), an albumin molecule has a globular spatial structure determined ultimately by the entropic properties of water, that is, by hydrophobic forces. The disposition of its links in this spatial structure also counts as information.

Biological evolution is irreversible and directed. It is difficult to calculate its entropic balance, but it is at least clear that its irreversibility and directedness are the result of export of entropy into the biosphere. We learned the basics of Darwin's theory from our school textbooks. That theory subsequently gained much new content from the successes of genetics and molecular biology. Of course, the raw material of the theory of evolution—the random mutations of genes and

[8] Author of *Problems of biological physics*, Moscow. Nauka, 1974.

[9] "When two or more amino acids combine to form a peptide, water is removed, and what remains of each amino acid is called an 'amino-acid residue'." *Trans.*

chromosomes—has no directedness to it. However, Darwin discovered the powerful agency directing the process of evolution, namely natural selection, whereby those mutants that are better adapted to their environment survive to produce more offspring. Natural selection forces an evaluation on organisms of wide genetic variability. This evaluation consists in the extraction of valuable information. We shall discuss this further in the next section.

Apart from the selection for adaptive change, there is a stabilizing selection process leading to the discarding and extinction of mutant strains differing markedly from the norm on either side in some significant way. This stabilizing selection also has a cost in energy and entropy. This might be called the "Red Queen's effect" from Lewis Carroll's *Through the looking glass*:

> "Well, in our country," said Alice, still panting a little, "you'd generally get to somewhere else—if you ran very fast for a long time, as we've been doing."
>
> "A slow sort of country!" said the Queen. "Now, here, you see, it takes all the running you can do, to keep in the same place. If you want to get somewhere else, you must run at least twice as fast as that!"

Thus natural selection gives direction to biological evolution. However, natural selection is neither the only nor the most important directing agency, although even some of the best textbooks speak of it alone in this respect. The words of Darwin himself are all too often forgotten:

> We clearly see that the nature of the conditions is of subordinate importance in comparison with the nature of the organism in determining each particular form of variation;—perhaps of not more importance than the nature of the spark, by which a mass of combustible matter is ignited, has in determining the nature of the flames."

Thus the structure and properties of populations of living beings—considered as dissipative systems—participating in the evolutionary process, are what mainly give direction to that process. The structure of an organism and the possible ways in which it can change, are determined by how it has evolved up to the present, and this is closely related to its ontogenesis, that is, to its individual development. For example, all land mammals have four limbs because they are descended from fish of the subclass *crossopterygii*[10], which had just four fins.

In this sense evolution is like a game of chess. If a game of chess has proceeded for a sufficient number of moves, it cannot then develop further as if those moves had been quite different. Like the evolutionary process, a chess game is irreversible, channelled, and directed. There's no going back.

In evolution, as in chess, there is a double limitation. Evolution makes unchanging use of nucleic acids and albumins, and the genetic code. In chess this corresponds to the fixed rules of play. The evolutionary process is restricted to one direction, and by what has gone before. A chess game is restricted according to the choice of opening gambit and the initial development of the pieces.

[10] "A subclass of bony fish including both fossil and living lobe-finned or tassel-finned fish".

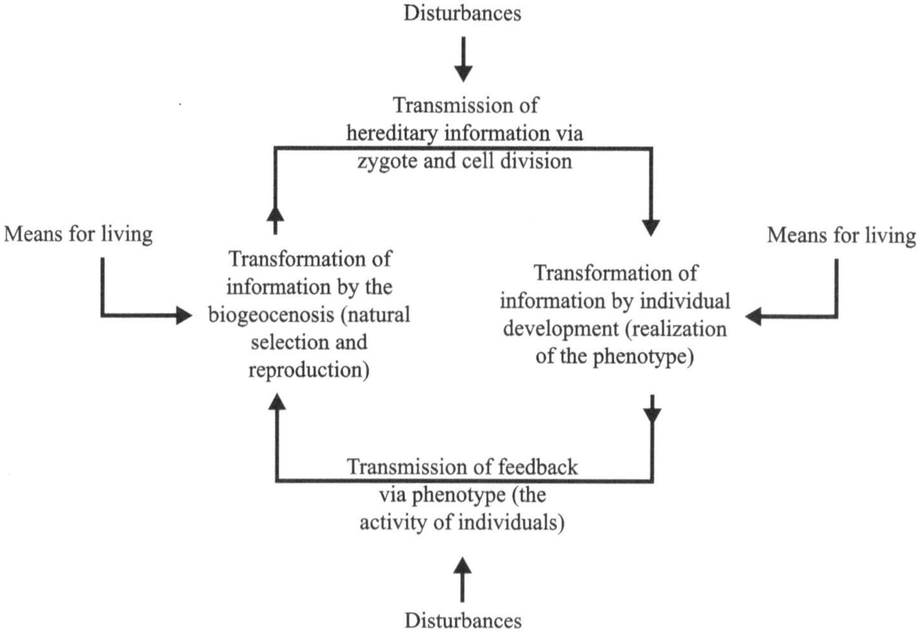

Figure 8.1: Scheme of direct and inverse communications in a biogeocenosis (natural selection and reproduction) according to Schmalhausen.

Chess is based ultimately on the creation, reception, and evaluation of information. In similar fashion, the biological information contained in individuals and populations is generated and "evaluated" via natural selection acting within imposed limits.

Of course, this is but a broad analogy. For one thing, in evolution there are not simply two opposing players in the game.[11] However, analogies are useful: they help to clarify concepts and in this way represent sources of new information.

One of our greatest biologists, I.I. Schmalhausen,[12] proposed a cybernetic, informational, formulation of the mutual relations between organisms and their environment in the course of the evolutionary process. This translation of evolutionary theory into the language of information theory turned out to be highly instructive.

The functioning of organisms takes place in "biogeocenoses", that is, in ecological systems formed by all the coexisting and interacting flora and fauna, together with the geological and climatic elements of the environment. The biogeo-

[11] Perhaps "a war of all against all"? *Trans.*

[12] Ivan Ivanovich Schmalhausen (1884–1963), Russian and Soviet zoologist and evolutionist.

cenosis receives information about the state of a given population—the totality of individuals of a single species in a particular location—by means of certain feedback connections, activating a specific mechanism for transforming the information into control signals and means for communicating these to the population. The full informational scheme is shown in Figure 8.1.

The hereditary information stored in the genes—in the molecules of DNA—is transmitted from a population to the succeeding one only after it has been transformed in the biogeocenosis. It is not the genes that multiply, but individuals, representing component parts of the biogeocenosis.[13] The mechanism of hereditary transmission of information is contained within the separate individuals, but the changes in the biogeocenosis affect the population as a whole. Information about these changes is transmitted to the population via the breeding of selected individuals, whose offspring is incorporated into the population. The population changes—evolves—but the information about this change is communicated via individual breeding.

We are using the term "information" here only in a qualitative sense, that is, not quantitatively. Nonetheless, its use is instructive. Here is a clear example.

Lamarck claimed that there are acquired characteristics that are inherited. We know that the distant ancestors of the giraffe had short necks. Lamarck thought that the evolution of the modern giraffe proceeded by a gradual lengthening of the neck. Mum and Dad stretch their necks to reach the leaves on tall trees, and this lengthening is passed on to their children.

However, this would mean that information about the elongation of the parents' necks had to be communicated from their neck cells to their reproductive cells—and, moreover, communicated in an effective manner: the reproductive cells would have to change precisely in such a way as to cause longer-necked offspring to be produced. None of Lamarck's followers even looked for a mechanism that might transmit the requisite information—and quite justifiably, since no such mechanism exists. An individual's acquired characteristics are never inherited. That this is so is made especially clear by the above informational analysis.

Schmalhausen's work *Questions in the cybernetics of biology* ends with some conclusions of great importance. He stresses that what is crucial in evolution is not the amount of information contained in a chromosome, cell, or individual, but the quality or value of that information: "In every case where a comparison and selection of information takes place, this proceeds on the basis of an evaluation in terms of quality... In biology, evaluation of the quality of information is of fundamental importance".

The value of biological information

It follows that we should look into the problem of the *value* of information as it applies to biology.

[13]Although it is widely accepted that biological evolution takes place at the level of the basic replicator—the gene. (See Richard Dawkins *The selfish gene.*) *Trans.*

Earlier we gave a definition of the value of information in terms of its irredundancy and indispensability. We shall now consider the reception of the genetic information encoded as a sequence of nucleotides in a DNA chain. This is a "four-letter text" representing a program for synthesizing albumins. The essence of molecular biology is expressed in the dictum "one gene—one albumin chain". The program is composed in the genetic code, according to which each codon, that is, subsequence of three nucleotides, corresponds to a particular amino acid residue. The number of possible such triples is equal to the number of arrangements of three letters from a four-letter alphabet, that is, $4^3 = 64$, while there are just 20 amino acid residues. Hence the genetic code is highly redundant; there are amino acid residues coded for by one, two, three, four, and even six codons. The reception of such genetic information takes place in the cell's system for synthesizing albumins. The initial quantity of information in a DNA chain consisting of n nucleotides is

$$I_0 = \log_2 4^n = 2n \text{ bits.}$$

The value of this information lies in its use in synthesizing albumins. Each codon, that is, triple of nucleotides, corresponds to just one amino acid, so that the 20-letter albumin text is written with $n/3$ amino-acid "letters". Hence the amount of information in a corresponding synthesized albumin chain is

$$I_1 = \log_2 20^{n/3} = 1.44n \text{ bits.}$$

In view of the redundancy of the genetic code, the quantity of information contained in the albumin chain is less than in the DNA chain. The redundancy at this level is given by

$$R_1 = 1 - \frac{I_1}{I_0} = 0.28.$$

At the next level, we take into consideration the fact that certain amino acids composing an albumin can be replaced by related ones without altering the properties of the albumin. In other words, many point mutations,[14] expressed by the substitution of a different amino acid for the correct one, turn out to be neutral from an evolutionary point of view. If among the 20 aminoacids there are, say, four interchangeable ones on average, then

$$I_2 = \log_2 16^{n/3} = 1.33n \text{ bits,}$$

and the redundancy at this level is

$$R_2 = 1 - \frac{I_2}{I_0} = 0.33.$$

At the fourth level, we take into consideration the possibility of substituting one albumin for another. For instance, in a number of cases, various proteolytic ferments or enzymes necessary to the functioning of digestive systems are mutually

[14]That is, mutations in just a single nucleotide "letter". *Trans.*

interchangeable. If the cell's synthesizing system encounters two different DNA sequences, consisting of n_1 and n_2 nucleotides respectively, then it may synthesize distinct albumins, one with $n_1/3$ amino-acid links, and the other with $n_2/3$. By way of example, let's suppose that $n_1 = 0.75n_2$. Then

$$\begin{aligned} I_0 &= 2n_1 + 2n_2 = 4.67n_1 \text{ bits}, \\ I_1 &= 1.44n_1 + 1.44n_2 = 3.36n_1 \text{ bits}, \\ I_2 &= 1.33n_1 + 1.33n_2 = 3.10n_1 \text{ bits}, \end{aligned}$$

and, if the albumins are interchangeable,

$$I_3 = 1.33n_1 \text{ bits}.$$

The values of the corresponding redundancies R_0, R_1, R_2, R_3 are then respectively $0, 0.28, 0.33, 0.72$, and the increasing sequence

$$\frac{I_0}{I_0}, \frac{I_0}{I_1}, \frac{I_0}{I_2}, \frac{I_0}{I_3}$$

of relative values works out to be

$$1, 1.40, 1.50, 3.51.$$

Alternatively, one can define the value of a codon as the degree to which the amino acid residue it codes for is irreplaceable. We now know the DNA "texts" of a great many albumins, including DNA texts of a variety of species that code for the albumin with the same function. For example, we know the structure of the hemoglobins[15] of many vertebrates. This makes it possible to trace the evolution of albumins. The more distant two biological species are from one another, the greater the difference in the number of "letters", and therefore amino-acid residues making up the respective albumins with the same function. Comparison of mutational substitutions of amino acids in such albumins allows one to determine which amino acids are more and which less irreplaceable. The more irreplaceable an amino acid is in an albumin, the greater its informational value.

How does the value of information—its degree of irredundancy or irreplaceability—change over the course of biological development, both evolutionary and individual?

Fundamental to the theory of evolution is the notion that all forms of life have evolved from a single common progenitor. All multicellular organisms ultimately stem from unicellular ones. The one and only illustration in Darwin's *The origin of species* depicts the scheme of divergence of species occupying different ecological niches (Figure 8.2). For example, as was shown by the Soviet biologist N.N. Vorontsov,[16] the ecological niches occupied in the Old World by various

[15] "The iron-containing oxygen-transport metalloprotein in the red blood cells of vertebrates." *Trans.*

[16] Nikolaĭ Vorontsov (1934–2000), Soviet zoologist, ecologist, and evolutionist. Former Soviet minister of the environment.

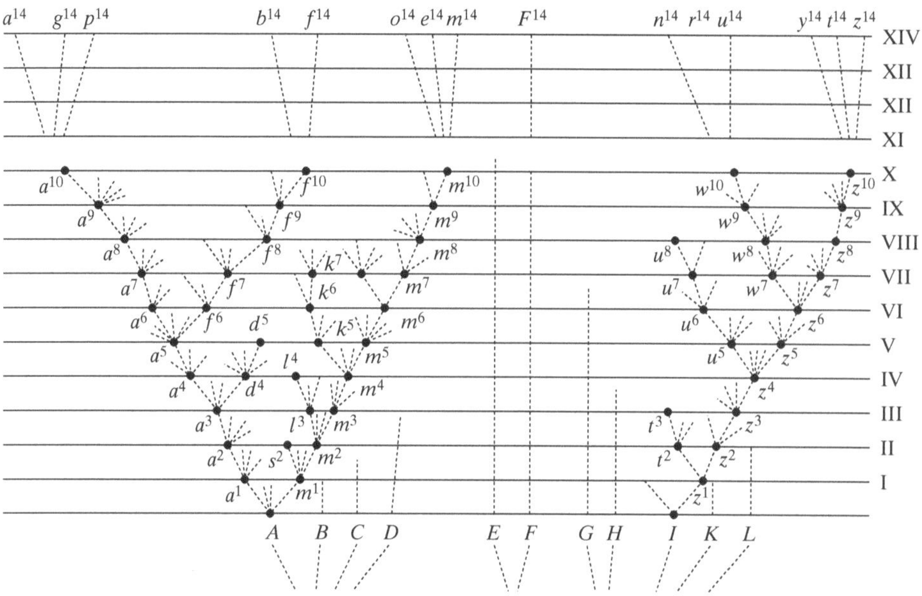

Figure 8.2: The scheme of divergence of species according to Darwin—the only diagram in *The origin of species.* The letters A, B, ..., L each represent common species of large genus. The fan of dotted lines of unequal lengths proceeding from A represents its varying offspring. The intervals between the horizontal lines each represent a thousand or more generations. The symbols a^1, m^1, etc. denote well marked varieties, and a^{10}, f^{10}, m^{10} new species. The symbol F^{14} denotes a species preserved without change.

species of rodents, were in the New World occupied by descendants of marsupials. Of these, some became tree-dwellers, others burrowed underground, and yet others made their habitats in rivers and lakes.

Clearly, such divergence represents a growth in irreplaceability, and so in informational value. Although members of different species do sometimes crossbreed, their offspring is invariably infertile. In this sense, therefore, biological species are irreplaceable.

An increase in irreplaceablity occurs also in individual development. At the beginning of the 20th century the German biologist Spemann[17] carried out his famous experiments in ontogeny. His subjects—or victims—were, as was usual, amphibians, chiefly newts. He showed that already at a very early stage of embryonic development, one can indicate the part of the embryo from which a particular organ will later develop. For example, it was shown that a certain part of an embryo during late blastulation or early gastrulation[18] would subsequently develop into an eye; this portion of the embryo, the rudiments of the eye, is termed a "presumptive" eye. Spemann grafted an early presumptive eye onto different places of a relatively mature embryo. The results were not especially interesting: grafted to an eye region, the presumptive eye grew into an eye, grafted to the ear region, it produced part of an ear, and in the kidney region, part of a kidney, and so on. However, when Spemann took the presumptive eye from a more developed embryo, at the stage of a neurula, then the presumptive eye turned out to be a "definitive" eye: it grew into an eye no matter which region of the embryo it was grafted onto. A diagram of these beautiful experiments is shown in Figure 8.3. As the biologists say—and biologists like specialized terminology—the "totipotency" of the region (initially constituting a presumptive eye) changed to "unipotency", that is, its ability to develop into any kind of tissue was at some stage lost, and subsequently it could only develop into one particular kind of tissue. Observe that this change marks an increase in the degree of irreplaceability, and so an increase in the value of the information contained in the molecules of the embryo, wherever and whenever they function so as to effect this chemical transformation.

In all organisms, from bacteria to human beings, we find the very important albumin cytochrome *c*, an indispensable participant in the so-called respiratory chain. The functioning of this chain of chemical transformations is accompanied by a process that might be called "charging up the battery", that is, synthesis of the substance ATP (see the first section above), which stores chemical energy, and ends with the oxidation of organic compounds by means of atmospheric oxygen, yielding the end products carbon dioxide CO_2 and water H_2O. Of course, like all biological systems, the functioning of the respiratory chain depends on the export of entropy.

We know the genetic "texts" of the cytochrome *c* of many species. Com-

[17] Hans Spemann (1869–1941), German embryologist. Nobel laureate in physiology or medicine for 1935.

[18] *Blastula*, *gastrula*, and *neurula* are terms used for the embryo at certain definite stages in its development.

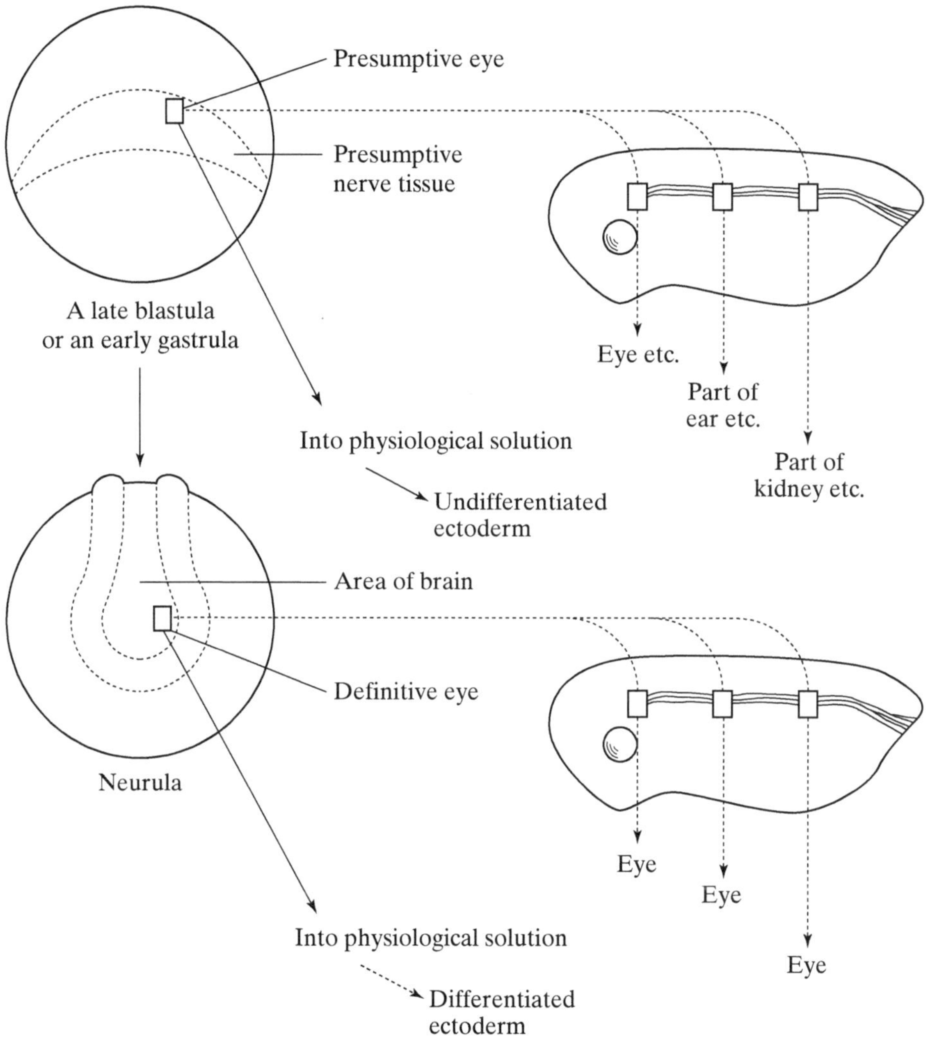

Figure 8.3: Grafting a presumptive and a definitive eye.

Table 7: Relative values of cytochrome c in conventional units.

Mammals		*Birds*	
Man	0.00	Penguin	0.00
Resus-macaque	-0.10	Chicken	-0.05
Donkey	-0.34	Duck	-0.30
Horse	-0.43	Emu	-0.30
Pig	-0.58	Pigeon	-0.58
Rabbit	-0.66		
Whale	-0.88		
Kangaroo	-0.88	*Reptiles*	
Dog	-1.06	Tortoise	-0.80
Elephant	-1.22		
Bat	-1.25		

parison of the values—that is, degrees of irreplaceability—of the corresponding amino-acid residues shows that among mammals, human cytochrome c has the greatest value. Among birds, the cytochrome c of penguins is of greatest value (just as Anatole France's *Penguin island*[19] would lead us to believe). The values of cytochrome c for some species, relative to that for human beings or penguins, are given in Table 7.

This table gives just differences in values, with the value of cytochrome c in human beings set at zero in the left-hand column, and that of penguins set at zero in the right-hand column.

What significance does this have? Of course, the value of cytochrome c has nothing whatever to do with human beings having—or not having—greater intelligence than other living creatures. What is behind this is that mankind has travelled a longer evolutionary path, as a result of which less valuable, that is, more readily replaceable, amino acids have been replaced by more valuable ones.[20] Incidentally, the more valuable amino acids are, generally speaking, also more rare.

Imagine a stamp-collector whose aim is not to increase the number of stamps in his collection, but to exchange less valuable stamps for more valuable ones. This is analogous to what happens with cytochrome c.

Thus it would seem that the value of the information contained in living organisms increases over the course of evolution.[21] Obviously, for this to be possible, the information existing originally[22] must have been of low value, that is, of high redundancy, so that replacements could occur tending to increase informational

[19] *L'Île des pingouins*, 1908.

[20] It is not clear here in what sense the author understands mankind as having undergone a longer evolutionary development than other creatures. *Trans.*

[21] This statement would seem to require some elaboration, for instance to account for the above differences in values of cytochrome c across species. *Trans.*

[22] At the dawn of life? *Trans.*

value.

In biological systems "trigger situations" are constantly being realized. These are situations where single signals carrying very small amounts of information have disproportionately large consequences. For example, a single molecule of colicin introduced into the cell of a bacterium is sufficient to kill it, and one prick with a needle is enough to induce an unfertilized frog ovum to divide, initiating ontogenetic development (Loeb's[23] experiments).

Trigger situations of this sort occur in ontogenesis, where they are necessary in connection with differentiation of cells and morphogenesis, that is, structure-formation. As noted earlier, triggering actions, which communicate information of especially high value, tend to occur in situations of great instability involving a large store of preparatory information.

In the course of individual or evolutionary development of an organism, selectivity of reception increases along with the increase in informational value. A biological system continues perfecting its ability to select or discriminate among the various kinds of information impinging on it. A frog reacts only to moving insects, and a bat, using ultrasound echo location, apprehends only reflected signals, and not direct ones. The choosing of what is of value in the information bombarding us from all sides is basic to the creative activity of human beings.

As we saw earlier (in equation (7.24)) the cost in energy of receiving one bit of information is at least $kT \ln 2$ joules. This is the minimal cost of one bit of *any* information, valuable or otherwise, irreplaceable or not. The selection of precisely the *valuable* information does not require any additional expenditure of energy. For this purpose, it suffices, for example, that the membrane of receptor cells have an appropriate structure. The cost in energy involved in the creation of this structure was met earlier on, during the preceding stages of biological evolution.

Thus a living system is capable of discriminating valuable from less valuable information, without incurring any special cost. Or, to put it differently, over the course of its development a living system approaches ever more closely to a state where it pays at least $kT \ln 2$ joules in energy for each bit of information of value to it, but wastes no energy apprehending information of no value to it.

Complexity and irreplaceability

In his last paper, published in the journal *Knowledge is power*, the pre-eminent Soviet biologist N.V. Timofeev-Resovskiĭ[24] wrote that, in order to construct a "theoretical biology", one will need a third law of evolution in addition to the two already established, namely the principles of natural selection and of covariant replication, the latter relating to DNA. The gene-stuff—the DNA—is duplicated, that is, copied, with each cell division. Here "covariance" means simply that any

[23] Jacques Loeb (1859–1924), German-born American physiologist and biologist.

[24] Nikolaĭ Vladimirovich Timofeev-Resovskiĭ (1900–1981), Soviet geneticist and evolutionary biologist.

mutations—that is, changes in the sequence of nucleotides of the DNA chain—are also duplicated.

As the third law of evolution one may take the principle of growth of complexity. Now clearly mammals are more complex than blue-green algae. It may seem rather obvious that complexity has risen over the course of evolution,[25] but what exactly do we mean by "complexity".

A rigorous definition of complexity enabling the concept to be quantified, is given in works of Kolmogorov, Chaitin, and Martin-Löf.[26] Any object may be encoded, using any agreed-upon system, as a sequence of zeros and ones. The *complexity* of the object is then the least length of such a binary sequence that will capture enough information about the object for it to be reconstructed, that is, decoded. In other words, the complexity of an object is the length expressed in bits, of the most economical program for generating a binary sequence describing the object.

Suppose we receive the two messages: 0101010101010101 and 011000101110 0101. Which is the more complex? The second, of course, since it is less ordered. The program that produced the first message, is telling us: $(01)^8$, or "write 01 eight times". The second message cannot be abbreviated in this way; any program specifying it would probably have to be as long as the message itself.

Here is another example. The number 3.14159... is communicated. An economical program would simply produce the symbol π, defined as the ratio of the circumference of a circle to its diameter.

It is noteworthy that in this definition complexity is equated with randomness: a random sequence of symbols, say binary digits, is assigned a degree of complexity coinciding approximately with the length of the sequence expressed in bits. Any minimal program is bound to be random rather than ordered in some way. For, consider a program P, minimal with respect to generating a sequence of binary digits comprising a message M. If P is not random, then by definition there must be a shorter program P′ that will generate P. However, then the message M can be generated by means of the algorithm: "Compute P from P′, and then M from P." Hence P could not have been minimal.

Consider the complexity of a biological system. In this context, of course, the concept is relative. For the biologist, the brain of a bull is a highly complex system, requiring perhaps millions of bits for a detailed specification, while for the butcher the description of the brain involves at most five bits, since it represents merely one of the roughly 30 parts of a bull's carcass destined for consumption, and $\log_2 32 = 5$.

The most complex objects in nature are living organisms, and among the most complex of these is man. We consider each person to be uniquely valuable, and therefore that human beings are not to be encoded by a program shorter

[25] At least for some organisms. *Trans.*

[26] Andreĭ Nikolaevich Kolmogorov (1903–1987), pre-eminent Soviet mathematician. Gregory John Chaitin (born 1947), Argentine-American mathematician and computer scientist. Per Erik Rutger Martin-Löf (born 1942), Swedish logician, philosopher, and mathematician.

than the actuality. In this sense, no "substitutes" are possible, we are each irreplaceable. This also applies to the creative output of individuals, for instance the artifacts of literature and art. It is impossible to devise a minimal program for *Anna Karenina*—there is no algorithm that simplifies a genuinely artistic work without sacrifice of its integrity.

In America and England one can find abridgments or synopses of Shakespeare's plays and Dickens' novels. After reading such summaries, a person can flaunt his "education", in spite of having acquired only a negligible fraction of the information contained in the full literary work. (We shall have more to say on this in the final section below.)

The objective taxonomy first drawn up by Carl Linnaeus[27] is fundamental to scientific biology. From the viewpoint of this hierarchical classification, each organism is not just an individual. For example, L.N. Tolstoĭ (Leo Tolstoy) is not only a great writer, he is also a representative of the kingdom of animals, of the phylum of chordates, of the order of vertebrates, of the class of mammals, of the genus of primates, of the species *Homo sapiens*, and, finally, of the variety *Homo sapiens sapiens* (presumed distinct from the variety *Homo sapiens neanderthalensis* of neanderthalers). Within each subdivision in the hierarchy, all representatives of that subdivision are considered equivalent, in the sense of having the same minimal set of characteristics defining the subdivision. In terms of such a classification, there are no "irreplaceables". As a representative of the species *Homo sapiens*, every person—regardless of differences of skin-color or other nonessential characteristics—is as good as any other.

Such a biological classification requires a minimal "program" or set of characteristics defining each subdivision or *taxon*. As one ascends the hierarchy through ever narrowing taxa, from kingdoms up to species, and, further, to an individual organism, these programs of specifications increase in length and complexity.

The scientific study of human beings starts, of course, from the level of the species, and introduces further subdivisions, always striving to minimize the respective programs. Thus the anthropologist and ethnographer classify the races and nationalities of human beings, the anatomists, physiologists and psychologists the actual manifestations of the phenotype—that is, the structure and functioning of the body, and personality types, such as the choleric, sanguine, depressive, and phlegmatic types.[28]

As we have seen, the concept of complexity is relative—it depends on the level of scrutiny, that is, of perception. At the level of the animal kingdom, a human being is equivalent to a fruit fly, at the phylum level to a lancelet,[29] at the order level to a porcupine, at the class level to a lemur, at the genus level

[27]Since Linnaeus drew up his taxonomy in the 18th century, it has been very substantially revised using better criteria—in particular, genetic— for the various groupings. *Trans.*

[28]This is the ancient-Greek typology of personalities based on the so-called four "humours": yellow bile, blood, black bile, and phlegm. Needless to say, in the world of modern science it has lost all credence. *Trans.*

[29]Any of several marine animals of the genus *Branchiostoma*. *Trans.*

to a chimpanzee, at the family level to *australopithecus*, and at the species level to *Homo erectus*.[30] In this sequence, complexity—the length of the program of specifications—increases

It might seem obvious that in the course of evolution complexity should increase. However, there are counterexamples to this suggestion. For example, when a life form adapts to a parasitic lifestyle, it usually undergoes simplification rather than increasing complexity, and certain animals (such as the amphibian *Proteus anguinus*, or white salamander) that took to living in caves eventually lost their eyes.

Here is an example where the male of a species has undergone extreme simplification in the course of evolution. The female of the marine worm *Bonellia viridis*, or green spoonworm, has a large (≈ 15 cm) sausage-shaped body, and is a complex organism capable of a variety of functions. The male, on the other hand, has a flat unpigmented body which grows to one or at most three millimeters in length, lives on or in the sexual ducts of the female, and is capable only of fertilization.

Such simplification is also encountered among vertebrates. In four families of deep-water anglerfish, special relations have evolved between the sexes. The male, much smaller than the female (in the case of *Ceratis hollbölli* the female is over a meter in length while the male is about 15 millimeters long!), penetrates the epidermis of the female. Once so ensconced, his jaws, eyes, and digestive system atrophy, so that he ultimately becomes merely an appendage for the production of sperm.

Thus as far as the idea that complexity increases with evolution is concerned, the situation is far from simple.

It is, however, obvious that the definition of complexity given above has much in common with the definition of value of a communication in terms of its irreplaceability. At the same time, the concept of value is the richer of the two, since complexity relates only to the communication as a whole, while separate components of it may be independently irreplaceable. Complexity characterizes the whole structure, while irreplaceability concerns functionality. It follows that over the course of evolution, although complexity may decrease in some situations, irreplaceability , that is, the value of information, always increases.

The growth of the value of information in the course of both individual and evolutionary development may be considered as one of the most important of the laws of theoretical biology. Of course, this law is not independent of natural selection, that is, it is derivative rather than standing alone.

---

[30]The genus *Australopithecus* was a kind of hominid widespread throughout Eastern and Northern Africa somewhere between three and four million years ago. *Homo erectus* lived in Africa about two million years ago, dispersing out of Africa into much of the Old World. *Trans.*

Complexity and Gödel's theorem

The problem of complexity also arises in connection with highly interesting and subtle mathematical questions, the investigation of which has led to important consequences of significance for science generally.

In 1931 the Austrian mathematician Kurt Gödel (1906–1978) proved his famous result on the incompleteness of arithmetic: he showed that no matter how one axiomatizes arithmetic (by means of a self-consistent, "recursive" system of axioms), there will always be statements in the language of formal arithmetic that are true but not provable from those axioms, that is, arithmetic is "incomplete".

The proof of Gödel's incompleteness theorem is related in spirit to the ancient paradox of Epimenides of Crete, who is said to have asserted that: "All Cretans are liars." One wonders if what he claims is true or false. If true, then what he says must be false since he is a Cretan; thus if true, then false, hence false. In fact the paradox may be reformulated in a stronger and simpler form, as follows: Consider the sentence "This sentence is false." Here we have a sentence that is neither true *nor* false. In effect, Gödel performed a similar trick, but with the criterion "true" replaced by "provable", and constructed from the sentence "this statement is not provable" an encoding into a statement in the formal language of arithmetic, which, consequently, was not provable, and therefore in fact true. The given axiom system might then be consistently supplemented with this statement, but then in this extended system, the same argument would yield true statements not provable within *it*.

Gödel's theorem has important consequences for science generally. The mathematician Yu.I. Manin[31] writes: "The successes of mathematics and areas of science that have been mathematized, led many deep thinkers to hope that there might be a few universal laws from which all other truths could be derived purely theoretically... Following on the work of Gödel, however, we can be sure that such hopes are groundless. Even apart from the question as to how complex the world might be, we now know that the deductive method is not sufficiently powerful. That method does not even suffice for deducing from a finite set of axioms all the true statements of arithmetic, formulated in the language of school algebra—such is the meaning of Gödel's theorem."

That is all very interesting, the reader protests, but what does it have to do with our discussion of the problem of complexity and irreplaceability?

The point for us is that, as a consequence of Gödel's theorem, it is in general impossible to prove the minimality of a given program that generates or encodes a sufficiently complex message. Take, for example, a statement like the following one: "The complexity of the sequence

$$0110110011011110001 0$$

[31] Yuriĭ Ivanovich Manin (born 1937), formerly a Soviet mathematician, now working in Germany and America. This quotation is from a paper of his in *Nature*, 12 (1975), a Soviet popular science magazine.

exceeds 15 bits." This calls for a proof that there is no shorter algorithm for this sequence than one requiring 15 bits to describe it. Such a proof would have to be formulated in the language of a formal system more complex than the sequence itself. The claim that the above sequence is random, for example, that is, cannot be encoded by fewer than the number of bits (20) comprising it, would settle the matter, but then this likewise would require a proof that might not in fact exist in the chosen formal system.

This does not mean that it is impossible to find a minimal program, but only that we may not be able to *prove* that it *is* minimal, and therefore also not be able to establish by means of logical argument its degree of irreplaceability, that is, the value of the information the message contains.

This conclusion is of general significance for science. The goal of science has always been that of finding a minimal program encoding and explaining the complex totality of facts being investigated. Already the mediaeval thinker Occam[32] asserted that in seeking the answer to a problem one should never multiply the number of basic concepts beyond the minimum necessary. For example, Newton's theory of gravitation explains at once the falling of an apple and the motion of the planets. However, Gödel's theorem shows that logic may be inadequate for establishing the minimality of a program for solving a problem. Intuition has always necessarily accompanied logical argument in scientific discoveries, and perhaps what we have just said can be viewed as a proof of the indispensability of intuition. L. I. Mandel′shtam[33] used to say that the fundamental equation of quantum mechanics—Schrödinger's equation—was guessed rather than deduced.

Intuition is the direct judgment of truth in the absence of logical argument.[34] Scientific creativity (and all the more so artistic creativity—see the next section) signifies the creation of new information, that is, the remembering of random choices—random, that is, intuitive, not provable by logical argument.

Information and artistic creation

We have mentioned before the universality and wide applicability of the concepts of information theory. We conclude this book by describing some informational approaches to the problems of artistic creation. The controversy reigning between, for instance, physicists and poets, makes no sense whatever. Science and art have far more in common than might appear, and in any case their resemblances are much more interesting than their differences.

The creation of a genuinely artistic work involves the creation of new information, since it involves the fixing, the committing to memory, of random choices.

[32] William of Occam, 14th century English logician and Franciscan friar. His principle is usually called "Occam's razor".

[33] Leonid Isaakovich Mandel′shtam (1879–1944), Soviet physicist.

[34] And unrestrained can lead, and has led, to wildly incorrect and potentially disastrous conclusions. *Trans.*

As in scientific work, the creative activity of an artist is determined by his own peculiarities and those of the society he lives in, but both he and the art experts who evaluate his work are largely unaware of these determining factors. And the greater the inspiration operating in the production of a work of art, the more they are hidden from consciousness. The logic of creative work often comes as a surprise to the artist himself. For example, A.S. Pushkin was amazed when his heroine Tatyana got married.[35]

A work of art represents an integral informational system. Thus the information conveyed by a poem is contained in all of its features: in the content, the vocabulary, the images, the meter, the euphony, the length, and the rhymes. It is impossible to separate these informational aspects of the poem, and in this sense a work of art resembles a living organism. It is created by a person in communication with the world about him, and so is itself a manifestation of life, the product of a creative mind. Once published, a poem acquires a life of its own as a non-isolated system retaining contact with its creator and entering into interactive relations with its readers and hearers. A poem allows of a real connection with the poet, alive in the verses—a connection of arbitrarily long duration. The wealth of emotions and ideas of Catullus,[36] or of the anonymous author of *The lay of Igor's campaign*[37] have remained for many centuries the property of their readers. The value of artistic information is of an aesthetic kind, being determined by the amount of influence it exerts on a receptor with the necessary preparation, that is, capable of apprehending the information, of feeling it deeply and imaginatively, and of evaluating it.

Artistic information—be it in the form of a poem, a picture, a sonata, or some other genre—is given directly to everyone, and everyone has the right to decide whether or not he likes it. While there's no arguing about taste, evaluations are another matter. One must certainly not use the false syllogism: "If I don't like it, it must be bad." It is very likely true that for a competent, serious evaluation of a work of art, one requires no less training than for competence in evaluating scientific work. What is needed is a "thesaurus".[38]

Reception of artistic information does not reduce to merely registering and storing it. Such reception involves a loss of part of the information in question, together with its enhancement. The first assertion is trivial and obvious: communication of information by any means is invariably accompanied by a loss of information through "noise". In the case of a work of art, the loss of information at reception is due to inevitable inadequacies of the consciousnesses of both author

---

[35] Aleksandr Sergeevich Pushkin (1799–1837), Russia's greatest classical poet. In his long poem *Eugene Onegin* the jaded hero Onegin first cooly rejects the shy confession of love of the youthful, intelligent, and beautiful Tatyana, daughter of a neighboring property-owner. When he encounters her many years later, respectably married, he realizes his mistake. *Trans.*

[36] Gaius Valerius Catullus (ca. 84 BC–ca. 54 BC), Roman poet.

[37] An anonymous epic poem written in Old East Slavic probably in the late 12th century. The poem recounts a raid in 1185, led by Igor Svyatoslavich, a prince of the Kievan Rus Empire, on the invading Mongolian Polovtsians in the region of the lower Don in 1185. *Trans.*

[38] See Figure 7.1, and the text immediately preceding it.

and receptor, that is, to unavoidable differences in their individualities. The artist has always felt keenly the impossibility of fully communicating his or her thoughts and feelings to the world. The very expression of those thoughts and feelings in verse form distorts them; as Tyutchev[39] put it:

... A thought uttered is a lie.

The increase in information on reception of a work of art is, on the other hand, far from being a trivial matter. The work of art activates, or programs, a stream of associations, thoughts, and feelings in the consciousness of the receptor, and stimulates the creation of new information by him. This information may simply remain in his mind uncommunicated, or be transmitted to those around him should he wish to, and be able to, share his experience of the work of art with them. This is the essence of the collaborative creative work of artist and receptor. Here the word "thesaurus", meaning store of necessary preparatory information, is not to be considered as meaning merely a sum total of bare facts recorded from what has been read, seen, or heard, like a card index, but is intended to embrace all of the intellectual and emotional richness of the receptor, including his ability to co-create. For this reason, an erudite person devoid of imagination and feeling may turn out to be a worse receptor than a less learned person capable of experiencing life vividly.

A necessary, but not sufficient, condition for artistic information to be valuable is that it be irreplaceable. Where is the value of an individual word higher—in an artistic or a scientific text? In an artistic text, of course—provided it is from a poetical or prose work of good quality, in which case not a single word can be changed without damaging the integrity of that informational system. Einstein wrote: "... There is not a single unnecessary note in Mozart's music." On the other hand, one and the same train of scientific ideas can be expressed in words equally well in a variety of ways.

It must, however, be emphasized that the notion of redundancy of information takes on a different meaning in the context of works of art. Unlike non-artistic texts—in newspapers, for example—, in artistic ones repetitions are far from being always redundant, that is, far from being devoid of fresh information. In ornamentation—of tiles or wallpaper, for example—a repeating pattern may have an emotional impact precisely because of the repetition. And this holds not only for applied art. A repeated refrain in a poem or in a passage of music has artistic significance. This shows again the importance of the integrity of an artistic work—the impossibility of delineating from it a rational content where repetition would indeed be redundant.

So what sort of information might be considered redundant, that is, valueless, in a work of art? Above all, cliché, banalities, and pointless repetition of what has already been fully gone over. Unlike an ordinary text, which is largly unaffected

[39]Fyodor Ivanovich Tyutchev (1803–1873), considered the last of the three great Russian Romantic poets, after A.S. Pushkin and M.Yu. Lermontov.

by redundancy, these types of redundancy mar a work with artistic pretensions, they destroy its integrity. A single stereotypical phrase is enough to render a whole poem worthless.

Of course, informational approaches will never be adequate for solving the fundamental problems of aesthetics. However, they allow a clearer understanding of those problems. To repeat: the value of a work of art is determined by the newly created information it contains. Generally speaking, the newer, the more unexpected that information is, the more valuable the work. Novelty is a necessary, though, of course, not a sufficient, condition of artistic value. Novelty, unexpectedness, can be expressed in different ways. I. A. Krylov's[40] fable *The comb* relates in simple folk language how a boy threw a comb into a river. But then:

Now naiads comb their hair with it.

Not mermaids, but naiads.[41] This prompts a complex of contrasting associations linking the Russian village with classical antiquity. This fairy tale was written during a period of fascination with classicism.

Spring, I come in from the street, where the poplar is amazed,
Where the distant landscape takes fright, where the house fears falling,
Where the air is blue like the bundle of linen
Of someone discharged from hospital.

Boris Pasternak[42]

Here, the contrasting comparisons and images make the description of spring especially informative. The reader feels spring, the sharp chill in the air, feels the poplar's "amazement" and the house's intoxication.

The new, irreplaceable information created by the artist has aesthetic value. In this regard, naturalistic painting is instructive. We quote S.V. Obraztsov:[43] "The absolute documentary resemblance between a person and his portrait, or a real pumpkin and a still life, may surprise, but this is not art, and indicates not talent, but merely long application and patience. When all is said and done, this is not so very difficult, and anyone capable of earning a grade of "excellent" for diligence can learn to do it."

This is absolutely correct for the simple reason that in painting a naturalistic picture the artist has set himself the task of producing an illusory encoding of reality, and the resulting work therefore contains only a minimal amount of new information. The only information created stems from the choice of a physical point of view of nature, the placement of a "frame" delineating the portion of the landscape, say, to be copied. The wax figures in Madame Tussaud's Wax

[40]Ivan Andreevich Krylov (1769–1844), Russia's best known fabulist. His early works were based on fables of Aesop and Jean de la Fontaine, but later ones were original with him. *Trans.*

[41]In ancient Greek mythology, a variety of water nymph.

[42]Boris Leonidovich Pasternak (1890–1960), Russian poet and novelist. Nobel laureate for literature in 1958.

[43]From *The relay race of the arts*, Moscow: Isskustvo (Art), 1978. This would appear to be Sergeĭ Vladimirovich Obraztsov (1901–ca. 1992), Russian artist and puppeteer, and founder of the Moscow Puppet Theater in 1931. *Trans.*

Museum are similar in this respect, that is, have roughly the same artistic content as a naturalistic painting. However, Madame Tussaud had no artistic pretensions in creating her wax figures. An aesthetic evaluation of a naturalistic painting is determined only by the object portrayed—the actual landscape, say—and not by any creative input by the artist.

"Informational aesthetics" is made especially striking by comparing an original artistic text with a translation of it into another language. An adequate translation of a single line of a poem is in general impossible since languages differ radically, and poetic information is contained in each and every word of the poem. Furthermore, the informational programs of the original text and of its translation are very likely to be different: they are intended for readers with different levels of reception perhaps, or different "thesauruses". A translated poem must stand as a poetic achievement in the language of the translation. The poet-translator has to solve the problem of optimally re-encoding the information contained in the original poem using words of another language. Different informational schemata may have differing values in poetry depending on the genre, style, or content. The translator has to determine the relative value of the various parts of the poem defining its essence, and preserve those of greater value while foregoing, if need be those of lesser import. The task is no less difficult than that of creating the original poem, and only a genuine artist is capable of performing it. It is not surprising that the greatest Russian poets—Pushkin, Lermontov, Tyutchev, and Blok[44]—made artistic translations.

Artistic information is created in the expectation that it will receive a full-valued appreciation by its receptors, and would not exist without this anticipation of its reception. The reception will vary over time and space. There was a time when the poems of Nekrasov[45] seemed to a wide circle of cultivated Russians (Pisarev and Plekhanov,[46] for instance) to be of greater artistic merit than those of Pushkin. Subsequently Nekrasov's work fell out of favor, and later came back into favor again, thanks in part to the influential poetical insights of Blok, which showed how much Russian poetry was indebted to Nekrasov. Such rises and falls in estimation represent sensitive registers of changing individual and social consciousness.

Throughout a person's life, he has new experiences, acquires new knowledge. His thesaurus changes continually, and with it his level of receptivity. That is why over the course of one's life one returns over and over to the great works of art: each time they seem different. Genius is unlimited informativity.

Granted that the creation of a work of art comes down to the creation of new

---

[44] Aleksandr Aleksandrovich Blok (1880–1821), considered to be one of the most gifted Russian poets since Pushkin. Leader of the Russian "symbolist" movement.

[45] Nikolaĭ Alekseevich Nekrasov (1821–1878), Russian poet, publisher, and critic.

[46] Dmitriĭ Ivanovich Pisarev (1840–1868), radical writer and social critic, one of those in the forefront of the democratic-revolutionary trend in Russia in the 1860s. Georgiĭ Valentinovich Plekhanov (1856–1918), Russian revolutionary and Marxist theoretician, a founder of the social-democratic movement in Russia.

information, that is, the remembering of random choices, to what extent do the artists themselves recognize this? We must give them the decisive word on this score.

In Karel Čapek's humorous story *The poet* he describes a sad scene: an automobile knocks down a drunken beggar, a woman, and races away from the scene of the accident without stopping. The few witnesses of the accident can recall neither the registration number of the vehicle nor even its color. One of these happens to be a poet by the name of Yaroslav Nerad. "When the accident occurred, he cried like a baby and ran home..." He also appears to remember nothing, but writes a strange poem, ending with the following lines:

> Passion was stilled. Lack of will... Oblivion.
> O swan's neck!
> O breast!
> O drum and those drumsticks—portent of tragedy!

The poet explains to the investigator that "Poetry is an internal reality. Poetry consists of free surrealistic images, born in the poet's subconscious... They are the visual and auditory associations that the reader must be interpenetrated with. Then he will understand."

And it becomes clear that "swan's neck" symbolizes the number 2, "breast" the number 3, and "drum and drumsticks" the number 5. The car with registration number 235 is found, and it turns out that indeed it was this car that killed the woman. Thus the poet had unconsciously and intuitively re-encoded the information "235" in the form of surrealistic verses. However, this represented not just a re-encoding, but the creation of new information concerning the poet's impressions. Here Čapek has in essence constructed a model—partly in jest, of course—of the creative process.

Intuition is needed in everything we do—in shipbuilding, in science, in everyday life—, but it is in art that it plays the leading role. In organizing collaborative creation with readers, viewers, or hearers, art "teaches inspiration", teaches intuition. This does not at all mean that there is no logic in art, that it does not "use algebra to verify harmony". In every work of art, logical, intuitive, and emotional elements co-exist. The parts these play may be variously unequal, as with the quantity and quality of the information created and communicated.

The relation between logic and intuition in art is the subject of Evgeniĭ L. Feinberg's substantive book *Cybernetics, logic, art.*[47]

The works of Leo Tolstoy provide many examples of interactions between the logical and the intuitive. He presents two faces to the reader: the artist-genius, intuitive, emotional, and direct, and, contrasting with this, the insistent preacher or teacher, using logical argument to advance his views of history, religion, science, and art.

---

[47]Published by *Radio i svyaz'* (Radio and communications), Moscow, 1981, and in English under the title *Art in a science-dominated world*, Gordon and Breach, 1987.

In the many-layered structure of the novel *Anna Karenina* one finds everything. There is the artistic development of the idea expressed by the epigraph at the beginning of the novel: "Vengeance is mine, and I will repay". There is another idea, of great importance to Tolstoy the preacher. The character of Levin expresses the author's ideology via the juxtaposition of his (Levin's) approach to life with that of Anna and her lover Vronskiĭ, and also her husband Karenin and brother Oblonskiĭ, all of whom exhibit selfish or ill-advised behavior.

However, the novel contains an antithesis to the approaches to life of all the leading figures, even Levin. This appears in the fifteen pages devoted to the artist Mikhaĭlov, which are among the best in the novel, although most readers pay them little attention. Here Tolstoy describes the artist, adverting on his psychology and the essence of his creative work. In doing this, he avoids all sermonizing and didacticism. Mikhaĭlov emerges as a genuine artist, a person of extreme emotions, resonating with every word, and understanding intuitively the essence of art. We are shown the recollection of a random choice in the following passage:

> The sheet of paper with the abandoned drawing was found, but it was smeared, and blotted with stearin. He nevertheless took the drawing, placed it in front of him on the table and, stepping back and screwing up his eyes, began to look at it.
>
> Suddenly he smiled and waved his hands in pleasure.
>
> "Yes, yes!" he said, and, taking a crayon, set to sketching rapidly. The spot of stearin had given the subject a new pose.

And further on:

> But in having made this adjustment, he had not changed the figure, but merely cleared away what had obscured it. It was as if he had removed covers preventing it from being fully visible...

And three pages later:

> He knew that it needed much care and attention to avoid spoiling the work by removing the cover, and in order to remove the cover completely...

Incidentally, the same idea had been expressed earlier in Baratynskiĭ's[48] poem "The sculptor", devoted to the myth of Galatea:[49]

> ...The unhurried, measured
> Chisel strips from the cherished goddess
> Layer after layer.

Baratynskiĭ and Tolstoy depict the creation of new artistic information as a coming to light of information that had been prefigured but hidden. It nonetheless still involves a free choice.

Tolstoy demonstrated the greatness of art and of the artist, who stands far above the main heros of the novel. In the brief Mikhaĭlov episode, Tolstoy the artist took full precedence over Tolstoy the preacher.

---

[48] Evgeniĭ Abramovich Baratynskiĭ(1800–1844), Russian poet, contemporary of Pushkin.

[49] One version of the ancient Greek myth has Pygmalion, king of Cyprus, sculpting a beautiful statue of a woman, falling in love with it, and praying to Aphrodite to bring the statue to life. Aphrodite obliged, thus creating Galatea. *Trans.*

In Mayakovskiĭ's[50] well known essay "How to make verses", he asserts that the writing of poetry is a difficult kind of consciously directed, productive labor: "Poetry begins at the point where a tendency arises." He is opposed to the Romantic tradition, and rejects inspiration and intuition. Poetry is "made" logically. He supports these claims by illustrating them with his poem "To Sergeĭ Esenin",[51] in which tragic, mordant humor, gloomy lyricism, the grotesque, and political declamation are all closely interwoven:

In this life
dying is easy.
To make a life
is significantly more difficult.

Mayakovskiĭ describes his conscious (not chance!) motivation for choosing almost every word of the poem. However, he also says "Novelty, novelty of subject matter and approach are essential to very poetical work."

In other words, the poet must "make" new information.

Mayakovskiĭ's essay also deals with "resonance and rhythm", which apparently come from nowhere. "Rhythm is the fundamental force, the fundamental energy of verse. It is impossible to explain it...".

In other words, writing poetry involves a random choice of rhythm, as well as words and images. The title of Mayakovskiĭ's poem "A cloud in trousers" was suggested to him by a chance conversation. He also talks of an image that occurred to him in a dream.

Of course, the thesis that writing poetry is the most difficult kind of productive labor, does not at all contradict its intuitive nature. Meyerhold[52] maintained that the suicides of Esenin and Mayakovskiĭ were the result of poor measures of worker-protection in the most dangerous area of production.

Long before Mayakovskiĭ, Edgar Allan Poe went much further in rejecting intuition and randomness in artistic creation. In his essay "Philosophy of composition", Poe tells of how he wrote his famous poem "The raven". He writes: "It is my design to render it manifest that no one particular point in its composition is referable either to accident or intuition—that the work proceeded step by

---

[50] Vladimir Vladimirovich Mayakovskiĭ (1893–1930), Russian poet of the revolution, and representative of early 20th century Russian futurism.

[51] Sergeĭ Aleksandrovich Esenin (1895–1925), Russian lyrical poet, mostly self-taught, and much-loved. The quoted excerpt from Mayakovskiĭ's poem would seem to be prompted by the last two lines of Esenin's brief poem of farewell, said to have been written in his own blood on the day of, or before, his suicide by hanging:

In this life there's nothing new in dying,
But on the other hand living's no newer.

Notwithstanding his veiled rebuke, Mayakovskiĭ also committed suicide, by shooting himself, whether out of unrequited love or disillusionment with the course the Soviet Union was taking under Stalin is not clear. Some conjecture that he was purged. *Trans.*

[52] Vsevolod Emil′evich Meĭerhol′d (1874–ca. 1940), Russian director, actor, and producer. Influential in modern theater through his experiments with physical being and symbolism.

step to its completion with the precision and rigid consequence of a mathematical problem."

Poe first considers the length of the poem. He reckons 100 lines as optimal. Yet the poem ended up having 108 lines.

The strongest impression is made by the poem's melancholy tonality, which is reinforced by the repetition of the "r" sound in the refrain. Poe chose the word "nevermore" to effect this. This word has necessarily to be repeated, but it is too monotone for a person to do so. Hence a bird rather than a person. But a parrot would certainly not work! Hence the raven, that croaks and symbolizes unhappiness.

Poe asks himself what the greatest melancholy is associated with. "Death was the obvious reply." Whose death? The death of a beautiful lady.

Then the problem arises of combining the lyrical hero mourning the death of his beloved and the raven repeating the word "nevermore". Hence the raven must answer the hero's questions.

Poe continues in this way with the aim of justifying his choice of the general structure of the poem—its length, division into stanzas, meter, and so on. Thus it would appear that all was calculated, planned in advance, developed logically, and that intuition, chance, and inspiration played no role in the poem's composition, so that no new information was created. The situation is paradoxical—after all, "The raven" is one of the finest works of the world's lyric poetry.

Poe is mistaken. He wrote his essay after "The raven" was published, as a polemic aimed at the maudlin romanticism of Longfellow and others. It suffices to read the poem again to be convinced that Poe's logic is born out of his poetic intuition, beginning with his choice of theme and ending with the poem's system of images, meter, and so on.

In Aleksandr Blok's lecture "On the poet's calling", dedicated to Pushkin and delivered just before his death, he says: "The poet is the child of harmony; and to him is vouchsafed a certain role in the world's culture. Three tasks are demanded of him: first, that of freeing sounds from the native elemental and unfathomable past where they abide; second, to impose harmony on these sounds, to give them form; and third, to convey this harmony to the outside world."

In other words, the poet must create poetical order out of primeval chaos, create artistic information. Science reveals the order actually existing in nature, by seeing through the external chaotic form of phenomena. In scientific work intuition is also indispensable—as evidence of which Gödel's theorem serves. Art is its own peculiar form of new order, "proving the unprovable", demonstrating the cogency of intuitive inference. Mozart and Salieri both created knowledge, but of the two, only Mozart created art.

* * *

We began this book with heat engines and ended it with a discussion of art. Although at first sight this association seems unnatural, in fact there is a direct path leading from Sadi Carnot to poetry, painting, and music. The main muse—not invented by man, but actually existing—is Entropy. Everything in which this world of ours differs from a grey, homogeneous chaos, arose and continues to exist as a result of the outflow of entropy into the surrounding medium. "Negative" entropy nourishes all of life and all that life has created, science and art in particular. Man produces negative entropy by creating new, irreplaceable information.

Index

MIX
Papier aus verantwortungsvollen Quellen
Paper from responsible sources
FSC® C105338

If you have any concerns about our products,
you can contact us on
ProductSafety@springernature.com

In case Publisher is established outside the EU,
the EU authorized representative is:
Springer Nature Customer Service Center GmbH
Europaplatz 3, 69115 Heidelberg, Germany

Printed by Libri Plureos GmbH
in Hamburg, Germany